D0947467

Public Meltdown
The Story of the Vermont Yankee
Nuclear Power Plant

RICHARD A. WATTS

The University of Vermont
CENTER FOR RESEARCH ON VERMONT

White River Press
Amherst, Massachusetts

Public Meltdown: the Story of the Vermont Yankee Nuclear Power Plant
Copyright© 2012 by Richard Watts. All rights reserved.

First published March 21, 2012

White River Press
PO Box 3561
Amherst, MA 01004
www.whiteriverpress.com

ISBN: 978-1-935052-60-9
eBook ISBN: 978-1-887043-02-1

Title Page Photo
Man walking on the Connecticut River with power lines and Vermont Yankee in background. Photo by Glenn Russell, *Burlington Free Press.*

Cover Photo
Senate President Pro Tem Peter Shumlin, D-Windham (center) and Senator John Campbell, D-Windsor, announce plans for a legislative vote on the future of Vermont Yankee. Photo by Glenn Russell, *Burlington Free Press.*

Cover Design: Susan McClellan, Irish Hill Design
Book Design: Sue Storey
Editorial Assistance: Jarred Cobb, Anne Bliss

Library of Congress Cataloging-in-Publication Data

Watts, Richard, 1959-
Public meltdown : the story of the Vermont Yankee Nuclear Power Plant / by Richard Watts.
p. cm.
ISBN 978-1-935052-60-9 (pbk. : alk. paper) -- ISBN 978-1-887043-02-1 (ebook)
1. Vermont Yankee Nuclear Power Station (Vernon, Vt.) 2. Entergy Corporation--Public relations. 3. Nuclear power plants--Vermont--Management. 4. Nuclear power plants--Licenses--Vermont. 5. Nuclear power plants--Risk assessment--Vermont. 6. Energy policy--Citizen participation--Vermont. 7. Environmental policy--Citizen participation--Vermont. I. Title.

TK1344.V47W38 2012
621.48'3097439--dc23
2012001217

Dedicated to my parents
Kathleen Heidi Watts
Simon Francis Watts

Contents

The Characters

Organizations

Central Vermont Public Service–investor-owned electric utility, one-time part-owner of Vermont Yankee, headquartered in Rutland, Vermont.

Citizens Awareness Network–citizen organization founded in 1991, opposed to Yankee Rowe (Massachusetts) and Vermont Yankee.

Department of Public Service (DPS)–state agency that represents consumers before the PSB and conducts state energy planning.

Entergy–diversified energy-holding company headquartered in New Orleans that manages nuclear power plants and provides electricity directly to customers.

Green Mountain Power–investor-owned electric utility headquartered in Colchester, Vermont. One-time part-owner of Vermont Yankee.

New England Coalition on Nuclear Pollution–organization founded in 1972, opposed to the continued operation of Vermont Yankee.

Nuclear Regulatory Commission (NRC)—Created as an independent federal agency in 1974, the NRC oversees the civilian nuclear power industry, including all aspects of radiological safety.

Public Service Board (PSB)–quasi-judicial state agency that oversees electric utility decision-making in Vermont.

Vermont Public Interest Research Group (VPIRG)–statewide environmental and consumer-advocacy organization founded in 1972.

Vermont Yankee–620-megawatt nuclear power plant located in Vernon, operating since 1972.

Individuals

Paul Burns–Executive Director, VPIRG

Brian Cosgrove–Vermont Government Affairs Director, Entergy

James Douglas (R-Middlebury)–Governor of Vermont, 2003-2010

Arnie Gundersen–Member, Vermont Yankee Public Oversight Panel

David O'Brien–Commissioner, Department of Public Service

Curt Hébert–Vice President of Corporate Relations, Entergy

Tony Klein (D-East Montpelier)–State Representative and Chair, Vermont House Energy and Natural Resources Committee

David Lamont–Director of Regulated Utility Planning, Department of Public Service

Sandra Levine–Senior Attorney, Conservation Law Foundation

J. Wayne Leonard–President and CEO, Entergy

James Moore–Energy Advocate, VPIRG

Duane Peterson–Board President, VPIRG

Mary Powell–CEO, Green Mountain Power

Ray Shadis–Technical Advisor, New England Coalition

Neil Sheehan–Public Affairs Officer, Nuclear Regulatory Commission

Peter Shumlin (D-Putney)–Governor of Vermont, 2011-present; Senate President, 2007-2010

Jay Thayer–Vice President of Nuclear Operations, Entergy

Robert Stannard–Lobbyist, Citizens Awareness Network

Robert Young–CEO, Central Vermont Public Service

Public Meltdown

ONE

Public Meltdown

O n March 11, 2011, a massive earthquake deep in the Pacific Ocean triggered a tsunami with waves up to 40 feet that tore into Japan's northeastern coast. Earthquake seismometers immediately sent signals to 11 area nuclear reactors. Even though the reactors rapidly powered down, heat continued to emanate from each reactor's core. At Fukushima, three reactors experienced the catastrophic physical accident known as a meltdown—when technical systems, often compounded by operator error, are unable to control the heat at the reactor's core. Electrical systems destroyed by the tsunami could not power the pumps that circulate cooling water. In each of the three most serious civilian nuclear accidents of the last 20 years—Three Mile Island (1979), Chernobyl (1987) and Fukushima (2011)—reactor cores reached meltdown, turning fuel rods into a molten mass and reaching temperatures of 4,000 degrees. Decisions by government and industry were clouded in secrecy and technological errors were compounded by human failures.[1]

This book is about another kind of meltdown, a public meltdown that took place over an eight-year period as Vermont citizens and political leaders became increasingly concerned about the management of a nuclear plant within state borders. Human errors and events outside of the control of the plant operator contributed to this meltdown. In a crisis environment, plant owners made a series of poor choices that undercut their credibility in Vermont and increased public concerns.

The nuclear plant at the center of this story has been a core piece of Vermont's electric load for 40 years, providing about one-third of the state's electricity since it first came on-line in 1972. Yet, in the 2000s, Vermont Yankee experienced a very public meltdown. The one-time fixture of the state's electric grid became a pariah, abandoned even by its former owners, the Vermont electric utilities. In 2010, the Vermont Senate voted to shutter the plant, making Vermont Yankee the first U.S. nuclear plant to be ordered closed in a public vote since Sacramento citizens terminated the Rancho Seco plant in 1989.

This is the story of a meltdown that has leaked its own poison into the atmosphere, infuriating state regulators, the public and legislative leaders and putting the state at odds with the federal government and with Entergy, the country's second-largest nuclear plant operator. This story also spotlights the relicensing of nuclear power plants and the role of state governments in the industry's efforts to prolong the life of the 40-year-old nuclear power plants located across the country. A *New York Times* reporter summarized it this way: "While Vermont Yankee demonstrates the ability of a single plant to squander public trust, it also demonstrates the emergence of the states in nuclear regulation."[2]

For its first 30 years, Vermont Yankee's primary owners were two regulated state electric utilities, managing the plant from their offices in Vermont. The cost of the plant's electricity was included in the utilities' rate base, with allowable costs set by state regulators. In 2002, the Louisiana-based Entergy Corporation purchased the plant, running it as a "merchant" facility and selling the electricity to the plant's former owners and regional wholesale markets.

Completed in 1972, Vermont Yankee is located in Vernon, a tiny New England town near Vermont's southern border with Massachusetts and across the Connecticut River from New Hampshire. Vernon was then and remains today a collection of a few thousand people (pop. 2,141) scattered across rich farmland in the Connecticut River Valley. Settled originally as a farming community, the town has more dirt roads than paved roads, a couple of churches, a school and a combined town hall and library.

Vermont Yankee is situated on 124 acres on a plateau at a sharp bend in the river about 1,000 feet from the town center. A half-mile

Figure 1. The Vermont Yankee nuclear power plant on a plateau above the Connecticut River. The plant's 22 cooling towers are in the foreground. The reactor is located in the multi-story rectangular building. The town of Vernon is in the background. Courtesy of Entergy.

south of the plant is a hydroelectric dam and a public boat landing. Canoeing and ice fishing in the river within a few paddles of the plant is not unusual. Like many places in Vermont, the area is sparsely populated. Although the plant is five miles from the shire town of Brattleboro, the region's merchant and business center, only about 35,000 people live in the 10-mile evacuation zone.[3]

The plant is one of the state's largest private employers, with more than 600 workers and a combined payroll of above $55 million. State and local taxes top $8 million and the plant purchases more than $6 million in local goods and services each year. As one of its former owners said: "For the first 30 years or so of its existence it was a quiet, well-run, valuable asset in the State of Vermont, providing roughly 25 to 30 percent of our power needs."[4]

Yet between 2002, when Entergy purchased the plant, and 2010, when the Vermont Senate voted to close the reactor, something happened. How did a plant originally seen as certain to be relicensed move from being a core piece of Vermont's infrastructure to rejection by Vermont's political leaders? What is the role of states that host nuclear power plants in the decision-making that affects their citizens?

This book presents the story of the public meltdown around Vermont Yankee, a story that spotlights the relations between nuclear plant owners and state officials, the interplay between the state and federal government and the roles of citizens and political leaders in these debates. Vermont Yankee, one of the country's oldest and smallest reactors, has a big story to tell.

The Dilemma of Aging Nuclear Power Plants

Vermont Yankee, like other U.S. nuclear power plants, was originally licensed to operate for 40 years. No new nuclear plant has been ordered in the U.S. since 1978 and the last new plant was completed in 1996.[5] To survive, the industry has turned to increasing thermal output from existing plants, running them more efficiently and seeking permits to add 20 years to their original 40-year licenses. Since 2000, the Nuclear Regulatory Commission—which has sole jurisdiction over radiological safety—has issued new licenses for all 71 of the reactors that have completed applications. An additional 15 applications are currently before the NRC, and another 17 plants have announced plans to seek license extensions. Conversations have started about "life beyond 60" as the 20-year license extensions will start to expire in the 2020 and 2030 decades.[6]

When Entergy purchased Vermont Yankee in 2002, the company took the unusual step of committing to apply for a state permit for a new license, signing an agreement with state regulators and testifying under oath to that commitment. Courts have upheld states' rights to review issues such as plant reliability, economics, environmental impact, land use, plant management and alignment with state energy plans.[7] Vermont's political and business leaders welcomed Entergy's purchase, citing the new owners' professional management skills, the benefits to Vermont and the state's continued oversight.

In 2006, legislators added a requirement that lawmakers also have a voice in the decision, requiring an affirmative vote of the legislature before state regulators could issue the permit for the additional 20 years. Entergy filed for a federal permit in 2006 and a state permit early in 2008, but the legislators did not vote in either 2008 or 2009. By the spring of 2010, relationships had deteriorated and Entergy was at odds with legislators, regulators and the public.

The meltdown was complete when one of Vermont Yankee's strongest supporters, State Senator Randy Brock (R-Franklin) stood up on the Senate floor and denounced Entergy: "We have a business partner in Entergy, that if its board of directors and its management were thoroughly infiltrated by anti-nuclear activists, I do not believe they could have done a better job destroying their own case. The dissembling, the prevarication, the lack of candor—have been striking."[8]

A few hours later, the Vermont Senate voted 26-4 against allowing state regulators to issue a permit, effectively requiring the plant to close in March 2012. A year after the vote, and 10 days after the tsunami devastated the Fukushima reactors, NRC commissioners approved the 20-year license extension. One month later, in April 2011, Entergy filed a lawsuit in federal court naming the state's governor, attorney general and regulatory body as defendants and seeking to have the Vermont decision overturned. A federal judge voided the Senate vote in early 2012, ruling that the Legislature had been motivated by radiological safety concerns—an area preempted under federal law. But the judge reaffirmed the role of state regulators, requiring a state permit for continued operation.

At the time of this writing, the case is now back before the Vermont Public Service Board, returning Entergy to where the company started in 2002.[9] An appeal of the judge's decision would put the state and Entergy back into federal court. Entergy remains politically and socially isolated, at odds with Vermont's governor and political leaders. State energy plans no longer include Vermont Yankee in the state's electricity future, and the plant's former owners now buy their electricity from elsewhere.

Other states, the nuclear industry and activists across the country continue to watch the unfolding story. In New York, the state's governor opposes the relicensing of two Entergy-owned reactors at Indian Point. In Massachusetts, a relicensing case for the Entergy-owned Pilgrim reactor is now in its sixth year. In New Jersey and Illinois, state regulators concerned with leaking radiation have increased their scrutiny of several reactors. In California, relicensing debates over several plants, including Diablo Canyon, are sure to stir continued questions about the role of states in the nuclear power debate.

Covering the Unfolding Story

During the eight years of Entergy's ownership, the story of Vermont Yankee grew from one centered near the plant in southern Vermont to a story of statewide interest. Unfolding events, coupled with the activities of citizens and environmental groups and Entergy missteps, catapulted the story forward. By the end of the decade, the Associated Press called the evolving story at Vermont Yankee one of the top state stories of the last 10 years, along with Governor Howard Dean's run for U.S. President.

Vermont Yankee's public meltdown also became a national story, covered by outlets from California to New York. By the time of the Senate's vote in February 2010, stories about Vermont Yankee had appeared in more than 100 national news outlets, including the *New York Times, Wall Street Journal, USA Today* and ABC News.[10] Would Vermont vote to reject a nuclear power plant? Would the state be allowed to continue to have oversight? Media outlets asked these questions as they followed the twists and turns in the story.

Vermont has a strong and vigorous public media culture, populated by numerous weekly and biweekly community-based newspapers, eight daily newspapers, at least three local television news channels and several active news radio services.[11] Over the years these outlets and the growing online community (e.g., VTDigger.org) have provided a comprehensive view of the evolving story. Reporters at just five Vermont media sources—the Associated Press, the *Burlington Free Press*, Vermont Public Radio, WCAX-TV and the *Brattleboro Reformer*— produced more than 1,400 separate news stories featuring Vermont Yankee between Entergy's purchase of the plant and the Vermont Senate vote. These five outlets provide a multi-medium, multi-dimensional view of the story of Entergy's public meltdown as told by some of Vermont's finest and most experienced reporters.

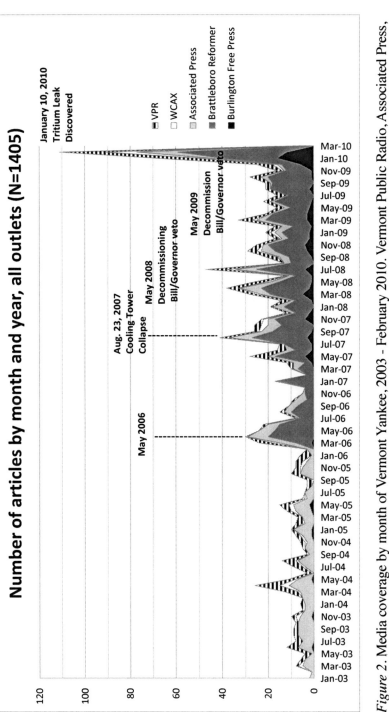

Figure 2. Media coverage by month of Vermont Yankee, 2003 - February 2010. Vermont Public Radio, Associated Press, *Burlington Free Press*, WCAX-TV, *Brattleboro Reformer* (*Reformer* starts in January, 2006).

Together, just five reporters at these outlets were responsible for about 60 percent of the bylined news stories.[12] These are very good reporters, reporters who provide the "first draft of history," telling the Vermont Yankee story—the AP's David Gram, VPR's John Dillon, the *Free Press*'s Terri Hallenbeck, WCAX's Kristin Carlson and the *Reformer*'s Bob Audette (who alone wrote more than 400 Vermont Yankee news stories).[13]

To tell the story of Vermont Yankee's public meltdown, I rely on their reporting, other press accounts, interviews with participants, reviews of the many public documents available and my own experiences as a citizen in Vermont during these years. Full citations of these sources, which include print and online media, can be found in the extensive footnotes. The following chapters focus on key moments in the evolving story, ending with a summary of the federal judge's decision in the lawsuit Entergy filed and the likely next steps. Legislators, activists, citizens, utility managers and public officials all play leading roles in this closely watched state drama, a drama which is still unfolding.

Notes

1. Stephanie Cooke, *In Mortal Hands: A cautionary history of the nuclear age* (New York: Bloomsbury, 2009); J. Samuel Walker, *Three Mile Island: A nuclear crisis in historical perspective* (Berkeley: University of California Press, 2004); Hiroko Tabuchi, "Japan panel cites failure in tsunami," *New York Times,* December 26, 2011.

2. Matthew L. Wald, "Edging back to nuclear power," *New York Times,* April 22, 2010.

3. "Nuclear neighbors: Population rises near US reactors," MSNBC, April 14, 2011, http://www.msnbc.msn.com/id/42555888/ns/us_news-life/.

4. Robert Young, in discussion with the author, August 24, 2011.

5. Robert J. Duffy, *Nuclear Politics in America: A history and theory of government regulation* (Lawrence, KS: University Press of Kansas, 1997).

6. David Lochbaum, in discussion with the author, December 20, 2011.

7. Michael Dworkin, in discussion with the author, January 19, 2012.

8. Center for Media and Democracy, "Under the dome: Vermont Senate vote on Vermont Yankee," February 24, 2010, http://www.cctv.org/watch-tv/programs/vermont-senate-vote-vermont-yankee.

9. Under PSB regulations, plants are permitted to operate while in the permit renewal process, even if the deadline for renewal has passed.

10. Matthew L. Wald, "Radiation levels cloud Vermont reactor's fate," *New York Times,* January 27, 2010; Rebecca Smith, "Vermont scuttles plans for reactor," *Wall Street Journal,* February 25, 2010; Matthew Mosk and Alice Gomstyn, "As Obama calls for more nuclear power, Vermont hits reverse," ABC News, February 24, 2010; Wendy Koch, "Vermont bucks Obama's nuclear call by voting to shut plant," *USA Today,* February 25, 2010.

11. William K. Porter and Stephen C. Terry, "The Media," *Vermont State Government since 1965,* ed. Michael Sherman (Burlington, Vermont: Center for Research on Vermont, 1999).

12. Based on analysis of 1,409 articles in news media database. John Dillon wrote 178 of 261 (68.2%) stories on Yankee for VPR. Susan Keese and Bob Kinzel contributed most of the remaining. David Gram wrote 80.5% of AP articles with a listed author. Kristin Carlson contributed 11 of 87 (12.6%) stories on Yankee for WCAX. Terri Hallenbeck wrote 66 of 104 (63.5%) articles on Yankee for the *Burlington Free Press.* Bob Audette wrote 441 of 526 (83.8%) articles on Yankee for the *Brattleboro Reformer* in the time period of the *Reformer* stories (1/1/07-3/1/10).

13. There are a number of reporters and media outlets not included in this analysis, notably Susan Smallheer at the *Rutland Herald,* several reporters in the *Rutland Herald/Times Argus* Montpelier Bureau, Stewart Ledbetter at WPTZ and *Seven Days.* Any mention of Vermont Yankee's media coverage in Vermont should also mention two excellent public affairs shows, Vermont Edition on VPR and the Mark Johnson Show on WDEV. For research reasons I chose the media outlets detailed here; a local newspaper, the state's largest paper, a wire service and a television and radio news broadcast.

TWO

Forty Years of History

Vermont Yankee was born in the heyday of nuclear power. In the 1960s and early 1970s, the peaceful use of nuclear power represented economic progress. "Too cheap to meter" nuclear power was to provide the fuel to grow the country's economic might, bringing jobs and growth. For much of the 1950s and 1960s, electricity growth was linear, increasing at more than seven percent annually. An engineer could lay a yardstick on a piece of paper and predict future growth.

At the same time, economies of scale and ever larger and more efficient (and centralized) sources of electricity reduced costs. Every year for almost two decades, electric use increased and the costs per unit of energy decreased.[1] Eisenhower's 1953 "Atoms for Peace" initiative envisioned nuclear power as contributing to this growth, building on the expertise developed in the U.S. military's nuclear weapons program. There was widespread public support and enthusiasm for nuclear technology in the Atomic Age. Walt Disney, for example, published a children's book, *Our Friend, the Atom*, in which the hero harnessed the power of the atom to bring peace to the world.[2]

In 1954, Congress created a new federal agency, the Atomic Energy Commission, to "develop, promote and regulate" the nuclear industry—for both commercial and military purposes. The peaceful use of the atom provided public cover for the continued development of nuclear weapons, bomb-grade materials and production facilities.

The two uses of nuclear power are intertwined. Commercial nuclear power, like nuclear weapons, requires enriched uranium. The spent fuel can also be reprocessed into plutonium, another material used to make nuclear bombs. The Atomic Energy Commission worked closely with Congress, industry and the White House to support nuclear research and development.[3]

Despite this support, electric utilities remained reluctant to embrace the new technology. One major stumbling block was the issue of insurance. Insurance companies were unwilling to insure the new power plants because of the potential risk of a catastrophic nuclear accident. Utilities refused to build nuclear power plants until the issue was resolved. In 1957, Congress approved the Price-Anderson Act, capping total liability in a nuclear accident at $500 million. Private insurance and plant operators provided $60 million, U.S. taxpayers were on the hook for the rest. In later years, Congress increased the cap to $750 million.[4]

The support of the federal government, ever-increasing electric demand and the subsidization of the first wave of power plants by manufacturers Westinghouse and General Electric led to a boom in nuclear plant construction. In the 10 years between 1965 and 1974, utilities ordered 223 nuclear reactors. Electric utilities ordered most of the U.S.'s present fleet of 104 nuclear reactors in the late 1960s and early 1970s during the great "bandwagon" market in nuclear power. The power plant at the center of this story, Vermont Yankee, started commercial operation at the peak of the nuclear power boom in 1972. In just three years between 1972 and 1974, 31 power plants, almost one-third of the nation's present fleet, received operating licenses.[5]

In 1973, the AEC envisioned more than 1,000 reactors in the U.S. by the year 2000. In the northeast, early nuclear reactors completed included Yankee Rowe in Massachusetts (1960), Indian Point 1 in New York (1962), Connecticut Yankee (1968), Oyster Creek, New Jersey (1969), Nine Mile Point 1 in New York (1969), Maine Yankee (1972), Vermont Yankee (1972), a second Massachusetts plant, Pilgrim (1973), and New York's FitzPatrick (1974), Indian Point 2 (1974) and Indian Point 3 (1976).[6]

The rapid expansion of nuclear power came to a halt in the mid-

1970s, due to a confluence of factors. The OPEC oil embargo forced sharp price hikes in fossil fuels, and oil-based construction costs skyrocketed, making nuclear power plant construction unaffordable. Borrowing costs also increased substantially at the same time, as electric utilities went deeply in debt to finance the new plants. Added safety regulations and other requirements also increased per plant costs. And in 1973, for the first time in 20 years, Americans used less electricity than they had the year before. Completed plants suddenly were selling electricity into a market glut and drawing added scrutiny from regulators. Regulators were increasingly unwilling to pass on plant costs to consumers. For example, regulators stopped allowing construction costs while the plant was being built to be included in the rate base.

These factors led to sharp increases in per-plant costs, which jumped from $600 per kWh in the 1970s to $3,500 per kWh in the 1980s. Cost overruns of 500 to 1,000 percent were common.[7] While plant costs for Vermont Yankee tripled from the original estimates, later plants experienced even more dramatic cost overruns. New Hampshire's Seabrook facility, 80 miles from Vermont Yankee, started operating in 1990 at a cost of more than $6.5 billion, bankrupting its parent company. During the plant's 17-year permit odyssey, plant costs ballooned 600 percent.[8]

Between 1975 and 1980, utilities ordered only 13 more reactors. The next decade saw many more canceled than ordered. The last new reactor ordered in the U.S. was in 1978, and the last one completed was Watts Bar in Tennessee in 1996.[9] Still, the wave of reactors ordered in the great bandwagon market of the late '60s and early '70s led to a sharp increase in the total number of plants. In 1960, there were two operating commercial reactors in the U.S., by 1985, there were 84.[10] Total reactors in the U.S. today number 104, close to one-quarter of the world's 2010 fleet of 435.[11]

In the face of increased public scrutiny, spurred by cost overruns and regulatory failures, the AEC's role as both regulator and chief promoter came under fire. In 1974, Congress reorganized the agency, renaming it the Nuclear Regulatory Commission and removing the nuclear weapons portfolio. Despite the change in name, the agency had the same structure and staff. By the time Ronald Reagan was elected in 1980, the agency was once again an advocate for nuclear power.[12]

While strong support continued for the industry at the federal level, opponents at state and regional levels were seeing increased success in adding safety and environmental protections and in delaying and even forcing the cancellation of new plants. The increasingly expensive upfront capital costs and long delivery times made the plants prohibitively expensive for electric utilities. Issues of land use, thermal pollution and the ability to safely evacuate large populations became hotly contested. In 1990, a Long Island utility sold the completed $5.5 billion Shoreham nuclear power plant to New York State for $1. Despite an NRC operating license, the plant never ran, although, in contrast to Seabrook, the utility owner survived.[13]

With new power plants increasingly difficult to build, the industry turned to increasing the operating capacity of the existing fleet, upping the thermal output of individual plants and extending the life of the plants beyond their initial 40-year licenses. During the 1970s and 1980s, nuclear plants ran only 60 to 70 percent of the time. By the 1990s, the industry had increased this number to above 90 percent. Increasing the thermal output of individual plants raised gross electrical output without building new plants. These changes meant that nuclear power continued to contribute about 20 percent of total U.S. electricity, even as electricity use increased and nuclear power plant numbers remained level.

However, with the licenses of almost half of the fleet set to expire by 2015, extending licenses became a critical issue for the survival of the industry. The industry and the NRC discussed this looming crisis starting in the 1980s, leading to the creation of relicensing regulations in 1991.[14] The NRC chose two plants— the Monticello pressurized water reactor in Minnesota and the Yankee Rowe boiling water reactor in Massachusetts—to pilot the relicensing process. The two-step process required plant operators to prove first that they were in compliance with their existing license, and then that the plant could operate safely for another 20 years.[15]

In the case of Yankee Rowe, NRC inspectors found possible issues with the embrittlement of the containment structure, a degradation of metal that could lead to radiation leaks. In 1991, the plant owners voluntarily closed the plant rather then make the needed investments. As with the Vermont Yankee plant (until 2002), a group of New England electric utilities owned and operated Yankee Rowe. The Monticello plant in Minnesota did receive a new license, but the process took many years. In 1995, the NRC amended the regulations to establish a more efficient, stable and predictable process. The new rule provided more credit for existing plant operator safety efforts, deemphasized the plant's previous operating record and restricted opposition groups' access to the process. Critics said the NRC had weakened the relicensing rules, because without the changes, older plants such as Yankee Rowe would not have been able to win license extensions.[16]

Since 2000, the NRC has issued new licenses for all 71 of the reactors that have completed applications. An additional 15 applications are currently before the NRC, and another 17 plants have announced plans to seek license extensions.[17]

Vermont Utilities Seek Permit to Build Vermont Yankee

Vermont utilities decided to join the nuclear boom at the height of the nuclear bandwagon. In 1965, utility orders for nuclear power plants were at an all-time high. Electricity demand was increasing an average of 7 percent annually, and prices were dropping. The heavy capital investment and potentially lower operating costs of nuclear power made it an attractive investment for utilities, which could incorporate the costs into their rate bases. Because rates for regulated utilities are based, in part, on total assets, increased capacity raises gross revenues, leading to greater profits. Utility groups in the New England region had banded together to raise the capital to build plants, starting with Yankee Rowe, which was completed in 1960.

Vermont utilities wanted to be part of this action. At Central Vermont Public Service, Vermont's largest electric utility, president Albert Cree saw nuclear power as key to his vision of expansion and economic growth. CVPS joined with Green Mountain Power— together the two served about three-quarters of the state—to raise the

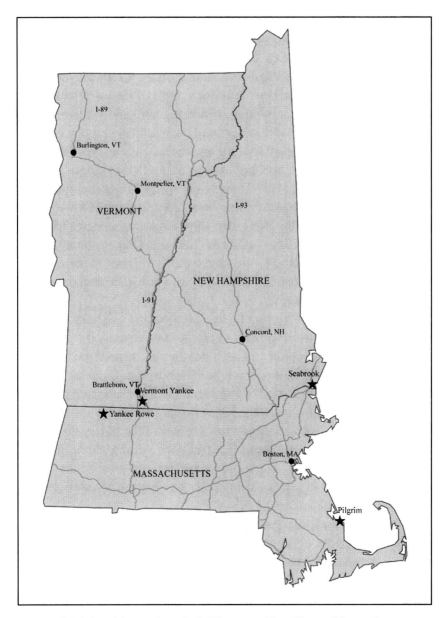

Figure 3. Major cities and roads in Vermont, New Hampshire and Massachusetts. Seabrook, Vermont Yankee, Yankee Rowe and Pilgrim nuclear power plants. Map created by Phoebe Spencer.

capital to build Vermont Yankee.[18] CVPS explored two sites: Vernon, on the banks of the Connecticut River in southern Vermont, and Orwell, on Lake Champlain on the western side of the state.

The decision to build Vermont Yankee in the 1960s played out against competing visions for Vermont's energy future—visions that pitted advocates for privately owned in-state power against advocates for public power. In Vermont, as in other rural places in the U.S., the high cost of serving rural areas had delayed the expansion of electricity. Senator George Aiken (R-Vermont) promoted the creation of rural electric co-ops and the development of hydroelectric facilities in upstate New York to provide low-cost electricity to the state. By 1962, two-thirds of Vermont's electricity flowed from hydro facilities on the St. Lawrence River—at $.08/ kWh, the cheapest electricity the state would ever see. In the early 1960s, other states demanded their share of the hydro power and Vermont's share declined.[19]

In 1963, Phil Hoff, Vermont's first Democratic governor, allied with municipal electric companies and the state's rural co-ops to seek new low-priced public power. The officials turned to Churchill Falls, a huge hydro complex being built in northern Canada. The Canadian owners committed to low rates and large amounts of electricity in return for assistance with project financing and the construction of new transmission lines to wheel the power through Vermont to markets in southern New England. Hoff and his allies promoted the creation of a new public authority to provide the electricity directly to state consumers, bypassing the privately owned electric utilities.

CVPS, Vermont's largest electric utility, and allied business groups opposed the proposed hydro deal, citing the uncertainty of working with a Canadian company and the costs involved in building the new transmission lines. Instead they argued for two new nuclear power plants, to be built in Vermont and owned, operated and controlled by Vermont-based companies. The electricity would be cheap, less than $.05/kWh and the in-state control would insure low rates for Vermonters, the utilities argued. The public power bill passed both the Vermont House and Senate but died in a conference committee negotiation. CVPS and GMP and the group of New

England utilities they led then moved to build Vermont Yankee, the site license from the AEC already in hand. The utilities first proposed the plant for completion in 1970 at a cost of $88 million. Vermont Yankee began operating in 1972 at a cost of $215 million, almost triple initial estimates. Electric unit costs, in this case per kWh, are set by state regulators and based on spreading plant costs across the customer base. State regulators initially set the rates at $.265/kWh, later dropping them to an average of $.20/kWh in the 1970s—four times higher than early projections.[20]

Vermont Yankee Design Raises Environmental Concerns

Vermont Yankee is a 620-megawatt boiling water reactor with a Mark 1 containment structure designed by General Electric. About one-third of U.S. nuclear plants today are boiling water; the others use a pressurized water system. A boiling water reactor relies on a closed-loop cooling system, with water circulated through the plant's core, heated by the nuclear reaction and turned to steam. Impurities are removed from the steam, which then drives turbines, creating electricity. The unused steam is condensed back into water and recirculated through the system. At Vermont Yankee, cooling is provided by water pumped from the Connecticut River, circulated through the plant, and returned to the river.

In the late 1960s and early 1970s, the biological impacts of returning heated water to the local water source had become a national issue, spotlighted by the growing environmental movement. The heated water would kill fish and destroy local marine habitats, opponents said. Charged with overseeing radiological safety, the Atomic Energy Commission refused to review or regulate the temperature of the discharged water.

Vermont Yankee became a flashpoint in this debate. Vermont citizens and political leaders as well as officials in neighboring New Hampshire and Massachusetts strongly opposed returning the heated water to the river. In Congress, Senator Edmund Muskie (D-Maine) took up the issue, hosting hearings in Vermont in 1966. Despite the attention, the AEC issued Vermont Yankee a permit in 1967, stating it had no authority over non-radiological issues. Pressure from political leaders continued, eventually leading to changes in the plant's design. The plant's owners installed 60-foot cooling towers from which waste heat is vented into the air as evaporated water and the remaining water is cooled through a system of fans and radiators before being returned to the river. Depending on the time of year and the plant's mode of operation, discharged water temperatures today are from 2 to 5 degrees Fahrenheit above river temperatures.[21]

Bowing to political pressure, the AEC later added water temperature issues to its site license permit and review process. However, the national debate on the warmed water brought added scrutiny to the industry, bringing in new critics and broadening the debate about nuclear power. The issue also cast a cloud over the openness and responsiveness of the AEC regulators. Increased costs associated with cooling the water also added to plant expense, further stirring state regulators and state and regional environmental officials.[22]

Vermont Yankee's containment structure also predates the containment domes that came in with later technologies and designs. In the early days of the nuclear power industry, plant designs evolved rapidly to incorporate operator experience and regulatory requirements. The Vermont Yankee reactor is housed in a square seven-story metal building. On the top floor, adjacent to the top of the reactor, is the spent fuel pool where more than 30 years of nuclear fuel is stored in a backyard-swimming-pool-sized body of water, maintained at a temperature of 85 degrees Fahrenheit. Starting in 2008, Vermont Yankee began storing used radioactive fuel outside of the pool in on-site concrete casks.[23]

In 1972, the AEC issued Vermont Yankee a 40-year operating license. The plant came on-line at exactly the wrong moment for new electricity generation in New England. The years of steady and predictable growth in electricity use had ended. In Vermont, peak electric load in 1974 was the same as it had been in 1970. As demand flattened, the energy crisis forced higher interest rates and increased fuel costs. Paying more to borrow money and build the plant, combined with the added expense of the cooling towers and a delayed opening due to faulty fuel rods, all drove up plant costs. In

Figure 4. The Vermont Yankee nuclear power plant. The spent-fuel pool and reactor are located on the top floor of the seven story building in the foreground. Storage for on-site radioactive waste is to the right of the reactor building. Photo taken from Hinsdale, New Hampshire. Courtesy of Entergy.

1974, CVPS and GMP had to request rate hikes; for GMP it would be the first such request in their history. [24]

By the end of the 1970s, the operating environment for the plant had stabilized. State consumers began using more electricity and the owners had resolved the initial problems with the plant. The plant became a core piece of Vermont's energy infrastructure, providing base-load electric power that met about 30 percent of the state's average annual power needs.

In the regulated electric utility world, the "allowable" costs of a power plant are charged to ratepayers. If the plant is expensive to operate, ratepayers are charged more, if operation is inexpensive, ratepayers are charged less. Nuclear plants require refueling outages about every 18 months (the plants are shut down while new fuel rods are inserted into the fuel assembly), and plant costs are higher during the outage years. Vermont Yankee met the industry's standards during the 1980s and 1990s, with one major 42-week outage that forced rates higher as electric utilities had to seek replacement power. During the 1990s for example, the plant ran on average 70 to 80 percent of the time. Costs of the power averaged between 5 and 6 cents per kWh depending on the year.[25]

Over time plant employment has averaged approximately 500 workers annually. Property tax payments and payroll taxes have been in the tens of millions and the plant contributes hundreds of thousands of dollars a year to local charities. Vermont Yankee is one of the largest private employers in the state, with employee paychecks integral to the local communities of southern Vermont.[26]

Opponents Seek to Close Vermont Yankee

Since Vermont utilities first applied for a permit in 1966, Vermont Yankee has had a strong base of opposition, centered near the plant in southern Vermont. In 1971, citizens in towns near the plant founded the New England Coalition on Nuclear Pollution. The Coalition joined other regional groups (e.g., the Clamshell Alliance) in bitter battles to oppose Vermont Yankee, New Hampshire's Seabrook plant and Yankee Rowe in neighboring Massachusetts. Over time, other local organizations joined in, such as Nuclear-Free Vermont and the Citizens Awareness Network, formed in 1991.

Opponents found the Atomic Energy Commission to be a difficult environment in which to challenge nuclear expansion programs in the late 1960s and early 1970s. The AEC limited public participation and interventions were highly technical, time consuming and costly. Opponents increasingly focused their advocacy at the state level and within various court systems. Issues related to emergency evacuation, water quality and the prohibitive expense of building the plants all became rallying points.

The early 1970s was also a time of tremendous growth in the environmental movement and a peak period in efforts to stop nuclear power. Although national groups, such as Greenpeace and the Natural Resource Defense Council, engaged directly, many of the battles were at the individual plant level.[27] The influential Union of Concerned Scientists, an organization that maintains a core group of highly skilled nuclear watchdogs today, formed in 1969 at the Massachusetts Institute of Technology. In 1972, Ralph Nader founded Critical Mass and a rally in New York City drew tens of thousands. There were massive protests at both ends of the country. Proposed nuclear power plants at Diablo Canyon in California and Seabrook in New Hampshire drew large waves of activists. In 1977, more than 2,000 people attended a rally at Seabrook; 1,400 were arrested and held for two weeks because they refused bail. Eventually, one of the two reactors planned for the site was canceled and cost overruns sent the owner into bankruptcy.[28]

In the early 1970s, Ralph Nader traveled to college campuses, inspiring the formation of a group of student-based organizations called Public Interest Research Groups (PIRGs). Students at Goddard College, Johnson State College, Castleton State College and the University of Vermont initiated the Vermont Public Interest Research Group (VPIRG), pooling their student fees and hiring lawyer Scott Skinner to set up an office in Montpelier, the state's capital. Too late to oppose the plant's initial licenses, these groups still fought nuclear power and Vermont Yankee at every opportunity. Capitalizing on legislative dismay at the higher-than-expected costs of Vermont Yankee, in 1975 the group advocated for a law adding legislative approval before any new nuclear power plant could be built. And, in 1977, legislators added a law requiring legislative approval of radioactive waste storage for any nuclear facility other than Vermont Yankee. [29]

The plant has also had a strong basis of support in the area, fueled by its large economic contributions, plant jobs and engagement in the local culture. The people who work at the plant live in and participate in their local communities. Vernon is strongly supportive of Vermont Yankee. Over the years, the town has benefited directly through local employment, tax revenues and charitable contributions.

But it's more than that. People who live in town work at the plant. In some cases they walk to work. Their kids attend the local school. It's been a part of the fabric of the town since the 1970s. As Mike Hebert, a member of the school board and a local legislator, told me one afternoon, "For those of us in town, we're very comfortable with the plant. But we're also very aware. And our neighbors, our friends, kids I coach on the Little League team, their parents work there." Hebert gestured towards a house across the road, "This one guy, next house over, lives across the street from the plant. Well, if it was dangerous and unhealthy, he wouldn't live there. So we're very comfortable in town. And I think we are very understanding of what goes on at the plant."[30]

Until Entergy purchased the plant in 2002, most of the debate was contained in Windham County in southern Vermont, away from the state's population centers and capital city. Public attention and press coverage from the rest of the state was minimal. While some political leaders criticized the plant over the years, efforts to close the plant never met with any success.[31]

"There were no major technical problems with the plant. We weren't in the headlines week after week. There was this sense that there was this quiet, well-run, valuable asset down on the Connecticut River," said Robert Young, former president of CVPS, one of the plant's original principal owners.[32]

Figure 5. The Vermont Yankee nuclear power plant is located in the town of Vernon. In this view looking to the southeast, the plant is framed in the foreground by the Miller Farm and some of the homes in town. Photo by Glenn Russell, *Burlington Free Press.*

Utilities Put Vermont Yankee Up for Sale

By the 1990s, the national environment for nuclear power plants had changed. As operating experience from existing nuclear power plants grew, the NRC continued to expand safety requirements. The accidents at both Three Mile Island and Chernobyl spurred extensive review. And, after the 9/11 terrorist attacks, the NRC required all nuclear power plants' owners to significantly upgrade their security systems.[33] Shortly before 9/11, the Vermont Yankee plant had scored dead last in a review of plant security by the NRC. The plant's owners had been relying on the local county sheriff to provide security, a business the sheriff operated in his spare time.[34]

In 1992, the utility owners of Yankee Rowe, the Massachusetts-based reactor a few miles from Vermont Yankee, shuttered that plant. The costs to decommission Yankee Rowe were estimated at $370 million—more than eight times the plant's original cost. In 1997, Maine Yankee, which had been completed at the same time as Vermont Yankee, was also closed.[35] Decommissioning costs there would eventually top $500 million. Both of these plants were owned by regional utilities, utilities generally much smaller than their counterparts in other states and regions. These utilities decided it was less expensive to close the plants and purchase the power elsewhere than to fund the upgrades needed to meet NRC safety requirements.

At the same time, utility restructuring (sometimes called deregulation) in the mid-1990s had started to change the vertical utility model, wherein companies owned power plants and provided retail electricity services directly to consumers. Instead, companies formed solely to own power plants, selling the electricity directly into wholesale markets. Nuclear power plants and other electric-generating facilities were increasingly managed by companies as for-profit entities, with the companies assuming the risks and selling the power at market prices. Firms such as Entergy that specialized in managing fleets of power plants were now common. These companies worked across state borders and were less subject to state regulation, with electric prices set by the market or the federal government, not state regulators.

Vermont Yankee's owners were becoming increasingly uncomfortable with the responsibility of owning and operating a nuclear power plant. Among other things, the utility owners were concerned about the costs of decommissioning the plant, which could be hundreds of millions of dollars.[36] Because the utilities owned Vermont Yankee, operating costs were charged to ratepayers. Vermont citizens were on the hook for operating failures and the costs associated with closing the plant. The new security rules were also a concern. Security upgrades starting in 2002 would cost more than $8 million.[37]

It was time, the Vermont utilities felt, to divest themselves of Vermont Yankee.[38] In 2001 they sought state permission to sell the plant to AmerGen, a nuclear power plant operator, for an initial price of $61 million, an amount which declined over time to under $10 million.[39] The deal needed the approval of the Vermont Public Service Board, the state agency charged with reviewing long-term power purchases, power plant purchases and the building of new plants to determine if they are in the "public good." In these quasi-judicial legal proceedings, other parties are allowed to intervene, question witnesses and file legal briefs.

Several groups in Vermont opposed the sale, most notably the Conservation Law Foundation, an environmental group that uses legal intervention to pursue changes in public policy. Lawyer Mark Sinclair, the former general counsel for the state's environmental agency, led the opposition. Sinclair blasted the agreement for its fire-sale price, arguing the plant was worth much more.[40] The Board agreed, requiring the Vermont utilities to competitively auction the plant, noting the concerns about loss of state control. Eventually, Entergy a large, diversified energy corporation, emerged as the winning bidder, purchasing the plant for $180 million. As part of the sale agreement, Entergy signed a 10-year electricity sale agreement with the plant's former owners and agreed to assume the costs of decommissioning, inheriting the $310 million already set aside for that purpose.[41]

Entergy Purchases Vermont Yankee

The contrast between the Vermont utility owners and the large, diversified Louisiana-based Entergy Corporation was striking.

Together Yankee's Vermont majority owners, CVPS and GMP, serve about 240,000 customers from their headquarters in Rutland and Colchester, Vermont, respectively. Combined gross sales for the two companies in 2010 were close to $600 million. Entergy, on the other hand, is a diversified energy company with gross sales above $11 billion and 15,000 employees spread across 20 states. The company owns merchant power plants, selling the electricity into wholesale electric markets, and also operates a regulated electricity business, providing direct electricity to more than 2.7 million customers in Arkansas, Mississippi, Louisiana and Texas.[42]

Based in New Orleans, Entergy Nuclear, a subsidiary of Entergy Corporation, is the second-largest nuclear generator in the U.S., operating 11 reactors in seven states. In the late 1990s and into the 2000s Entergy went on a nuclear power plant purchasing spree, buying the Pilgrim Nuclear Power Station in Plymouth, Massachusetts (1999), FitzPatrick (2000), Indian Point Units 2 and 3 in Westchester County, New York (2001), Vermont Yankee (2002) and the Palisades Nuclear Plant in Covert, Michigan (2007). All of these plants are merchant power plants not regulated at the state level, selling electricity into wholesale markets or directly to utilities. Together, Entergy's nuclear plants generated approximately 30,000 megawatts of electricity in 2009, roughly equivalent to the entire New England electric load.[43]

Entergy's scale and professional nuclear power operator experience enabled it to manage plants more efficiently. Nuclear expert teams could travel from plant to plant, meeting the extremely demanding and specialized skill needs of the industry. Two of Entergy's plants in the region, FitzPatrick, which started operating in 1974, and Pilgrim (1972) are also General Electric boiling water reactors, very similar to Vermont Yankee. Managing the fleet of nuclear power plants, including several of the same vintage, potentially allowed Entergy to reduce costs and increase the operating capacity of all the plants.

Entergy directly addressed two critical issues during the sale negotiations, issues that had been raised by citizens in public hearings and by opposition groups in the AmerGen case and were reiterated by then-PSB Chairman Michael Dworkin from the bench:

loss of state control and loss of economic benefits through selling the plant to an out-of-state entity.

To address these concerns, Entergy agreed to submit their planned license-extension request to Vermont regulators, signing a memorandum of understanding to that effect. The agreement stated that the "the board has jurisdiction under current law to grant or deny approval of operation ... beyond March 21, 2012." While state oversight of radiological safety is preempted under the 1954 Atomic Energy Act, a 1983 U.S. Supreme Court ruling maintained states' traditional oversight rights in other areas related to nuclear power, including environmental impacts, economics and reliability. In the agreement, Entergy also waived any claim to contest the deal based on federal preemption issues.[44]

Secondly, Entergy agreed to provide several key economic benefits to the state. Most importantly, Entergy negotiated 10-year power purchase agreements with the Vermont utilities, providing electricity at less than what they would have expected to pay had they retained ownership and thus reducing electric rates for most Vermonters.[45] The initial price was $.04/kWh and increased slightly each year, hitting $.042/kWh in 2010.[46] During the same period, equivalent electric prices on the wholesale market ranged from $.043 to $.088/kWh.[47] Entergy also agreed to a revenue-sharing provision with Vermont utilities (and ultimately with Vermont ratepayers), agreeing to share half of the revenue above $.061/kWh (indexed for inflation) for 10 years after the plant was relicensed, starting in 2012. Ratepayers would then share in any profits that Entergy would make if the plant's operating permit was extended. Entergy also agreed to a "low-market adjustor" under which Vermont ratepayers would not be penalized if market prices dropped during the initial 10-year contract period.[48] Furthermore, Vermont ratepayers would no longer be responsible if plant operations faltered or cleanup costs increased, a major concern of the utility owners.[49]

The Vermont Public Service Board, Vermont utilities and state policy leaders welcomed the sale, citing Entergy's professional management skills and the favorable terms of the agreement. The three-member Board pointed out that Entergy had resources and experience that exceeded those of the Vermont utility owners. And, the Board said the agreement Entergy had negotiated with Vermont utilities was "highly likely to produce an economic benefit for Vermont ratepayers."[50]

Vermont's Governor Howard Dean (D-Burlington) supported the agreement, as did the state's largest electric utilities. "After four years this is really a very happy day for Vermonters," said Robert Young, president of CVPS, the company that owned 35 percent of the plant. The state's top public advocate, Public Service Department Commissioner Christine Salembier, added that Vermonters "clearly win" in the deal.[51]

Not everyone, however, saw the deal in a positive light. "It is a sad day for Vermont ratepayers," said the Conservation Law Foundation's Sinclair. "From now on the future of the plant will be dictated by the bottom line financials of an out-of-state company."[52] The New England Coalition and VPIRG also denounced the sale.

Would Vermont Yankee be Licensed for Another 20 Years?

Between 2002 and 2009, Entergy invested more than $300 million in the plant, increasing its thermal output, building on-site storage for nuclear waste and preparing for relicensing. In January 2006, Entergy lawyers delivered a box of documents to the NRC. The 1,100-page permit application had required 40,000 engineering hours. The NRC permit application process would take more than five years and cost in excess of $10 million.[53] But with annual gross revenues of $200 million, meeting NRC relicensing requirements would be worth the investment.[54]

Following Entergy's submission, in March 2006, the NRC announced a 60-day public comment period. During the comment period, interveners could specify particular issues, or "contentions," that the NRC should consider. In 1995, the NRC had made it more difficult for interveners in relicensing cases, raising the burden of proof and limiting the issues to be considered. For example, issues of security, radioactive waste disposal or emergency evacuations were not considered germane. Ray Shadis, the New England Coalition's technical advisor, pointed out that Entergy had been working on the application for years, while the Coalition had only 60 days to review the documents, decide which issues to contest, find legal basis for those contentions and file responses.[55]

Still, Shadis and the Coalition did file to intervene and the NRC accepted three contentions as issues that should be considered. The contentions included technical issues related to the age and strain on essential plant components such as the steam dryer. In October 2007, the judicial arm of the NRC, the Atomic Safety Licensing Board, convened a public hearing in Brattleboro on the three issues that the Coalition had successfully injected into the license debate. One reporter called it a "David and Goliath" battle, pitting the New England Coalition, then deeply in debt, against Entergy and the NRC.[56] Four months after the hearing, in February 2008, NRC staff recommended the permit application be approved. The staff report is a critical step in the permit process, signaling likely approval by the NRC commissioners.[57]

Less than a month after receiving the NRC's staff endorsement, Entergy filed for a permit with Vermont state regulators as they had agreed to do when they purchased the plant. Entergy also would need an affirmative vote by the Legislature—based on the 2006 law known as Act 160—before the PSB could issue the license extension permit. (Under the peculiar way Act 160 was written, a lack of action by legislators also meant no permit could be issued).[58] During 2008 and 2009 relations between Entergy and legislators deteriorated. This came to a head in January 2010 when the company announced leaking radioactive fluids from underground pipes at the plant, pipes company officials had previously said did not exist. A month later the Vermont Senate voted 26-4 to close the plant. The very public meltdown of Entergy had dominated headlines and conversation across the state for almost two months.

The story took another twist when Entergy sued state leaders in federal court a month after receiving an NRC permit to operate another 20 years. The judge's decision, while voiding the Senate's vote, returned the case to the Public Service Board, putting Entergy back where they had started in 2002.

Table 1. **Key events in Entergy's Vermont Yankee history.**

June 2002............Entergy Nuclear purchase of Vermont Yankee completed.

February 2003.....Entergy requests 20 percent power increase.

April 2005...........Act 74 passes Vermont Legislature, approving on-site radioactive waste storage until 2012 and the creation of the Clean Energy Development Fund.

January 2006.......Entergy files a 20-year license extension with the NRC, seeking to operate the plant to 2032.

May 2006.............Governor Douglas signs Act 160 into law, requiring an affirmative legislative vote before the PSB can issue a certificate of public good.

June 2006.............Nuclear Regulatory Commission license extension hearing in Brattleboro.

March 2007..........NRC staff recommend approval of license-extension request.

March 2008..........Entergy files license-extension request with Public Service Board.

February 2010.....Vermont Senate votes to close the plant.

March 2011..........NRC commissioners approve license extension.

April 2011............Entergy sues Vermont.

January 2012.......U.S. District Court Judge J. Garvan Murtha issues decision striking down Act 160 but reaffirming the Vermont PSB's oversight role.

Notes

1 U.S. Energy Information Administration, "The changing structure of the electric power industry: An update," December 1996, pp. 108-109; also Entergy Corporation, "Entergy Vermont Yankee issues document," March 6, 2009.

2 Cooke, *In Mortal Hands*.

3 See Duffy, *Nuclear Politics in America*, and Cooke, *In Mortal Hands*, for a really good overview of the relationships between nuclear weapons, proliferation and nuclear power; also Rick Eckstein, *Nuclear Power and Social Power* (Philadelphia: Temple University Press, 1997).

4 See Eckstein, *Nuclear Power and Social Power*; Cooke, *In Mortal Hands*; and Duffy, *Nuclear Politics in America*.

5 NRC, "Information digest, 2011-2012," August 2011, Appendix A and B, http://www.nrc.gov/reading-rm/doc-collections/nuregs/staff/sr1350/v23/sr1350v23. pdf.

6 Ibid.

7 Duffy, *Nuclear Politics in America*, p. 175; Charles D. Ferguson, *Nuclear Energy: What everyone needs to know* (New York: Oxford University Press, 2011).

8 Eckstein, *Nuclear Power and Social Power*; Public Service of New Hampshire, "Our history," http://www.psnh.com/CompanyInformation/Our-History.aspx. Note: Interestingly, it is from this reactor that several of Vermont's electric utilities bought power to replace that from Vermont Yankee after the 2010 vote.

9 Duffy, *Nuclear Politics in America*.

10 NRC, "Information digest, 2011-2012," August 2011, Appendix A and B, http://www.nrc.gov/reading-rm/doc-collections/nuregs/staff/sr1350/v23/sr1350v23. pdf.

11 International Atomic Energy Agency, "Number of reactors in operation worldwide," http://www.iaea.org/cgi-bin/db.page.pl/pris.oprconst.htm.

12 Duffy, *Nuclear Politics in America*; Eckstein, *Nuclear Power and Social Power*; and Cooke, *In Mortal Hands*.

13 For a fascinating comparison of Seabrook and Shoreham, both completed in 1990, see Eckstein, *Nuclear Power and Social Power*. On Shoreham, a definitive read is an academic book, *Licensed to Kill? The Nuclear Regulatory Commission and the Shoreham Nuclear Power Plant* (Pittsburgh: University of Pittsburgh Press, 1997).

14 NRC, "Title 10: Code of federal regulations, Part 54: Requirements for renewal of operating licenses for nuclear power plants," 1991, http://www.nrc.gov/reading-rm/doc-collections/cfr/part054/.

15 David Lochbaum, in discussion with the author, December 20, 2010.

16 NRC, "Fact sheet on reactor license renewal," August 8, 2011, http://www.nrc.gov/reading-rm/doc-collections/fact-sheets/fs-reactor-license-renewal. html; Duffy, *Nuclear Politics in America*; Ferguson, *Nuclear Energy*; Cooke, *In Mortal Hands*; also Tom Zeller, Jr., "Nuclear Regulatory Commission changed nuclear relicensing rules," *Huffington Post*, May 9, 2011, http://www.huffingtonpost.com/2011/05/09/nuclear-regulatory-commission-changed-relicensing-rules_n_859692.html.

17 Nuclear Regulatory Commission; Nuclear Reactors Information Digest. http://www.nrc.gov/reading-rm/doc-collections/nuregs/staff/sr1350/v23/sr1350v23-

sec-3.pdf . Nuclear Energy Institute; License renewal, http://www.nei.org/resourcesandstats/nuclear_statistics/licenserenewal/.

[18] Vermont Department of Public Service, *Biennial Report July 1, 2000 – June 30, 2004*, May 9, 2005, http://publicservice.vermont.gov/index/05-09-05%20 FINAL%20DRAFT-AAA.pdf. Note: GMP has about 88,000 customers and CVPS 148,000. *Vermont Business Magazine*, 2010.

[19] Green Mountain Power, *Getting power to the people: The first 100 years of Green Mountain Power* (South Burlington, VT: Green Mountain Power, 1993).

[20] Ibid. Note: Peter Bradford summarized the costs to Vermont ratepayers of this decision at above $1 billion (1997 Aiken Lecture at the University of Vermont). For a detailed history of these events and the battle between public and private power advocates in Vermont see this excellent history by Lee Webb, "A history of electric utility regulation in Vermont: 1880-1965," (MA thesis, Goddard College), 1974; also see summary compiled by former Vermont State Representative Dean Corren (P-Burlington) at http://www.vce.org/LoosePages/repoweringvt.html.

[21] Vermont Agency of Natural Resources, "Amended discharge permit," Permit 3-1199, March 30, 2006, http://www.epa.gov/region1/npdes/permits/finalvt0000264permit.pdf.

[22] Duffy, *Nuclear Politics in America;* Green Mountain Power, *Getting Power to the People*.

[23] Entergy website, "EntergyFACTS," http://www.entergy.com/about_entergy/entergy_facts.aspx; "Entergy Vermont Yankee issues document," pp. 58-59.

[24] Green Mountain Power, *The Moon on a String* (Colchester, Vermont: GMP, 1983).

[25] Chris Dutton, in discussion with the author, May 27, 2011.

[26] "Entergy Vermont Yankee charitable contributions program supported more than 100 local community nonprofit organizations in 2010," *VTDigger*, February 3, 2011, http://vtdigger.org/2011/02/03/entergy-vermont-yankee-charitable-contributions-program-supported-more-than-100-local-community-non-profit-organizations-in-2010/.

[27] Eckstein, *Nuclear Power and Social Power*.

[28] The Vermont Electric Co-op was also forced into bankruptcy because of their part-ownership in Seabrook.

[29] Scott Skinner, "VPIRG's beginnings," *VPIRG Update*, Fall, 1992. Note: Skinner was the organization's first executive director, hired in 1972.

[30] Michael Hebert, in discussion with the author, June 10, 2011.

[31] David Gram, "Vermont Yankee nuke plant's critics still at it, 34 years later," Associated Press, October 26, 2006; Robert Young, in discussion with the author, August 24, 2011.

[32] Robert Young, in discussion with the author, August 24, 2011.

[33] NRC, "Backgrounder—Nuclear security," http://www.nrc.gov/reading-rm/doc-collections/fact-sheets/security-enhancements.html.

[34] "Residents question Vermont Yankee security," Associated Press. March 19, 2003. "Yankee boosts security," WCAX, November 8, 2002, http://www.wcax.com/story/1004469/yankee-boosts-security. Kathryn Casa, "Prue or false: The rise and fall of Windham Country Sheriff Sheila Prue," *Vermont Guardian*, June 30, 2006, http://www.vermontguardian.com/local/062006/PrueorFalse.shtml.

35 David Lochbaum, in discussion with the author, December 20, 2010; Union of Concerned Scientists, "Maine Yankee," http://www.ucsusa.org/assets/documents/nuclear_power/maine-yankee.pdf. Note: In this report, a company official tells the Maine regulators: "I think we did not keep up with the state of the art and we were too isolated from, maybe, from what was going on in the rest of the country." The plant's owners voted to close the plant permanently six months later.

36 Chris Dutton, Robert Young and Steve Terry, in discussion with the author, May 27, 2011, August 24, 2011, and April 18, 2011, respectively; see also CVPS's press release following the sale: Central Vermont Public Service, "CV lauds Vermont Yankee sale, consumer cost savings," August 15, 2001, http://www.cvps.com/aboutus/news/viewStory.aspx?story_id=138.

37 "Yankee boosts security," WCAX, November 8, 2002, http://www.wcax.com/story/1004469/yankee-boosts-security.

38 Chris Dutton, Robert Young and Steve Terry, in discussion with the author, May 27, 2011, August 24, 2011, and April 18, 2011, respectively.

39 State of Vermont Public Service Board, "Order dismissing petition," Docket No. 6300, February 14, 2001, http://www.state.vt.us/psb/orders/2001/files/6300fnl.pdf. Note: The actual sale price was net of transaction costs and time penalties, declining to an actual price of $8 million as the Board prepared to issue their order. That order was never issued as AmerGen withdrew their petition. Michael Dworkin, in discussion with the author, January 20, 2012.

40 Vermont PSB, "Investigation into general order no. 45 notice filed by Vermont Yankee Nuclear Power Corporation re: proposed sale of Vermont Yankee Nuclear Power Station and related transactions," Docket No. 6300, January 21, 2000, http://www.state.vt.us/psb/document/6300scopeschedorder3.pdf; Mark Sinclair, in discussion with the author, June 1, 2011.

41 Susan Smallheer, "Yankee sale gets approved," *Rutland Herald*, June 13, 2002; Entergy, "Annual report for 2002," March 17, 2003, http://files.shareholder.com/downloads/ETR/1610695088x0xS65984-03-199/65984/filing.pdf. Note: Plant actually purchased by Entergy Nuclear Vermont Yankee, a subsidiary of Entergy Nuclear, which in turn is a subsidiary of Entergy. Entergy owns a controlling interest in ENVY. While noting these ownership distinctions, in this book I refer to the plant's owner as Entergy.

42 Entergy, "EntergyFACTS," http://www.entergy.com/about_entergy/entergy_facts.aspx.

43 Entergy, "Company History," http://www.entergy.com/about_entergy/history8.aspx; "Entergy Vermont Yankee issues document;" Gordon van Welie, "2010 ISO/RTO metric report presentation," February 12, 2011, http://www.iso-ne.com/pubs/pubcomm/pres_spchs/2011/final_isone_metric_1_11.pdf.

44 Michael Dworkin, in discussion with the author, January 20, 2012; *Pacific Gas & Elec. Co. v. State Energy Resources Conservation and Development Comm'n*, 461 U.S. 190 (1983); Vermont PSB, "Memorandum of understanding among Entergy Nuclear Vermont Yankee, LLC, Vermont Yankee Nuclear Power Corporation, Central Vermont Public Service Corporation, Green Mountain Power Corporation, and the Vermont Department of Public Service," Docket No. 6546, March 4, 2002, http://www.leg.state.vt.us/jfo/envy/6545%20MOU.pdf.

45 CVPS, "Vermont Yankee sale to Entergy completed," July 31, 2002, http://www.cvps.com/aboutus/news/viewStory.aspx?story_id=120.

46 Vermont PSB, "Investigation into general order no. 45," June 13, 2002.

47 US EIA, "Wholesale market data," 2002-2011, http://www.eia.gov/electricity/wholesale/index.cfm.

48 Vermont PSB, "Investigation into general order no. 45," June 13, 2002, p. 71.

49 Ibid, p. 4; Chris Dutton, May 27, 2011.

50 Vermont PSB, "Vermont Public Service Board finds Entergy qualified to own and operate Vermont Yankee; approves sale and associated contracts," June 13, 2002, http://www.state.vt.us/psb/document/7404VT_Yankee_Reorg/Exhibit%20GMP-CVPS%20NRB-2.pdf. Note: Contracted parties were the Vermont Yankee Nuclear Power Corporation (VYNPC), which included at the time of the sale Central Vermont Public Service Corp (31.3%), New England Power (22.5%), Green Mountain Power (17.9%), Connecticut Light and Power (9.5%), Central Maine Power (4%), Public Service Company of New Hampshire (4%), Burlington Electric Department (3.6%), Cambridge Electric Light (2.5%), Western Massachusetts Electric (2.5%), Vermont Electric Cooperative (1%), Washington Electric Cooperative (0.6%) and Village of Lyndonville Electric Department (0.6%).

51 Sue Robinson, "Yankee sold to Entergy," *Burlington Free Press*, August 1, 2002.

52 Ibid.

53 David Gram, "Standards high for fighting Vermont Yankee relicensure," Associated Press, March 24, 2006.

54 Entergy official Jay Thayer told a reporter that Entergy spent "close to" $300 million upgrading the plant's systems between purchase and 2009. Susan Smallheer, "Entergy says it will give its 'best offer,'" *Rutland Herald*, December 16, 2009.

55 David Gram, "Standards high for fighting Vermont Yankee relicensure," Associated Press, March 24, 2006; Vermont Public Radio, "NRC opens hearings on Vermont Yankee's license extension bid," June 7, 2006.

56 Sam Hemingway, "Fed hearings shine light on anti-nuke coalition," *Burlington Free Press*, July 27, 2008.

57 "Entergy Vermont Yankee issues document." Note: Still under consideration were the three contentions raised by the NEC, now under review by the Advisory Committee on Nuclear Safeguards.

58 One key part of Act 160 reads: "(2) No nuclear energy generating plant within this state may be operated beyond the date permitted in any certificate of public good granted pursuant to this title, including any certificate in force as of January 1, 2006, unless the general assembly approves and determines that the operation will promote the general welfare, and until the public service board issues a certificate of public good under this section. If the general assembly has not acted under this subsection by July 1, 2008, the board may commence proceedings under this section and under 10 V.S.A. chapter 157, relating to the storage of radioactive material, but may not issue a final order or certificate of public good until the general assembly determines that operation will promote the general welfare and grants approval for that operation, 30 V.S.A. § 248, http://www.leg.state.vt.us/statutes/fullsection.cfm?Title=30&Chapter=005&Section=00248.

THREE

Cracks in the Wall

E ntergy, the new owner of Vermont Yankee, stepped into a state that has a strong independent streak. Vermont is also a place where community matters, where civic participation is common and where outsiders are not always trusted—not because of who they are, but because of the values they represent.

Vermonters' focus on self-sufficiency and self-reliance dates back to the early 1700s. Before Vermont joined the union as the 14th state, it was its own independent country, fighting battles with the British, annexing towns in neighboring New Hampshire and New York, coining its own money and negotiating a treaty with Canada. This independent spirit has manifested in many ways over the centuries. In 1941, three months before Pearl Harbor, Vermont declared war on Germany.[1] Thirty years later, Senator George Aiken (R-Vermont) famously proposed that the U.S. should declare victory in the Vietnam War and go home. In 2001, Senator James Jeffords (R-Vermont) left his party because it had become out of step with the values of his native state. Jeffords was later replaced by Senator Bernie Sanders (I-Vermont), the only independent in the U.S. Senate.

Combined with Vermonters' independent spirit is their commitment to individual freedom, with these values incorporated into the state's constitution. Vermont's early leaders wrote the first state constitution in the country to outlaw slavery. More than 200 years later, Vermont state legislators approved the nation's first civil union law, giving same-sex couples the same rights as heterosexual couples.

Vermonters' early drive for self-reliance ran headlong into the state's challenging climate—long, cold, dark winters, closed in by mountains and forests and rugged soil. It took a community to survive, a community made up of friends and neighbors. And this sense of community is integrated into how Vermonters help each other, forming a thread through Vermont's 246 separate incorporated towns and villages. These twin characteristics of independence and community are enshrined in the state's motto, "Freedom and Unity," and embedded in the personality and culture of the state.

The state's geography has influenced its development. The Green Mountain Range—mountains of granite—form a barrier north to south in the middle of the state. To the west is the fertile Champlain Valley and Lake Champlain, the sixth-largest freshwater lake in the U.S. To the east are hills sloping down to the Connecticut River, which forms a political boundary with neighboring New Hampshire. Vermont's 625,000 people are clustered in the Champlain Valley and in towns along the Connecticut River Valley and at the base of the mountains.

The state's "green" reputation stretches back many years. Environmental concerns cross party lines and the relationship between economic development and environmental protection is strong. The state's working landscape, populated by about 1,000 mostly small-scale dairy farms, provides scenic vistas for residents and visitors while providing local jobs and supporting local communities. Hillside development of cheaply built ski condominiums in the late 1960s spurred the passage of one of the country's first development-control laws. In 1970, Governor Deane Davis (R-Barre) devoted his entire State of the State address to promoting the new law—that came to be known as Act 250—to restrict untrammeled growth. Vermont is one of a handful of states that have banned billboards, and in 1972 became the second state, after Oregon, to pass a bottle deposit law.

Although citizens of a solidly "blue" state today, Vermonters are not necessarily anti-nuclear power. Public opinion polls have shown state voters to be mixed on this issue. Polls frequently show a strong minority opposed and a strong minority in support with a neutral-to-undecided middle. For example, a survey taken in January 2010, at the height of the public crisis around Vermont Yankee, found 60 percent of Vermonters in support of nuclear power and 30 percent opposed.[2]

Vermont citizens have a long history of public participation in civic matters, with the state ranking near the top of the country in the engagement of its citizenry.[3] Part of this stems from the localness of government in Vermont: the 246 small towns and cities were originally known as "little republics." Each of these republics has its own government, where citizens make decisions. It's been estimated that more than 9,000 Vermonters serve in elected or appointed public offices, offices that range from town lister to planning board member to state legislator.[4]

Vermont is home to Town Meeting Day, an annual gathering of local residents to debate town budgets and policies. For more than 200 years, Vermonters have been gathering on or around the first Tuesday in March, trudging through snow and cold to chat with their neighbors, snack on local foods and discuss town matters. The state's citizens care about and are deeply engaged in the issues that affect them. On average 20 percent of a town's registered voters turn out every year to spend a morning or an evening talking about their town face to face with neighbors.[5]

Starting in the 1970s, town meeting became a site for debating larger societal issues, ranging from the impeachment of President Richard M. Nixon to bans on genetically modified food organisms. In 1982, 160 Vermont towns voted to support a "nuclear freeze"—requesting the U.S. government to stop adding to nuclear weapon stockpiles. The nuclear freeze concept spread to other states and eventually reached the U.S. Congress. Senator Dan Quayle (R-Indiana), a member of the U.S. Armed Services Committee (and later U.S. vice president), told a reporter that "some grassroots caucus taking place in Vermont" is not going to decide U.S. military strategy.[6] Ten years later, Bill Clinton adopted the "town meeting" as a campaign strategy, staging town-meeting-like events in states across the U.S. to garner support in his quest for the presidency.

In addition to their independence, the state's people have a taciturn streak and a distrust of the values of outsiders. It's not where they are from, as Vermonters are an accepting group, it's

Figure 6. "Reddy Kilowatt" and Central Vermont Public Service workers celebrating wiring the last town in Vermont, Victory-Granby, in 1963. Courtesy of Central Vermont Public Service.

the values outsiders can represent. "Flatlander," a term sometimes used to describe non-Vermonters, becomes code for "incompetent, impractical, fleeting and unreliable."[7]

Entergy Comes to Vermont

Entergy's purchase of Vermont Yankee put them into a select group of companies producing and distributing electricity in Vermont. The state's history is interwoven with that of the electric companies that first powered the small towns and woolen mills, dating back to the 1800s. Electricity in Vermont, like the Vermont utilities that provide it, is part of the fabric of the state. The companies that make, distribute and manage electricity are key players in Vermont's political, economic and social structures. The advent of electricity is recent history in the more rural parts of the state, wired in the 1950s and 1960s, with utility poles extended to the last Vermont town, Victory, in 1963 (see *Figure 6*).

Electricity is much more than a commodity in Vermont, and Vermonters have a strong interest in how it is made and the companies that make it. Vermonters care about electricity because the state can be very cold and dark, but also because of their strong interest and engagement in social and economic issues. Thousands of Vermonters participated in debates about long-term contracts for hydro-power from the giant Canadian company Hydro-Québec in the 1980s. Many Vermonters participated in discussions about the planned construction of new natural-gas plants, wind turbines and high-voltage transmission lines in the 1990s and 2000s. State leaders led the national conversation about electric energy efficiency in the 1990s, creating the first statewide utility in the U.S. focused solely on conservation and efficiency (Efficiency Vermont). This decentralized, household-based approach requires the participation of thousands to succeed. Today Vermont vies with California in per capita spending on programs to reduce energy use.

Electricity is a key part of the state's economic and political culture. The decisions around electricity are very public. As longtime utility executive and former newspaper reporter Steve Terry said, "More than any other commodity that I know, electricity is imbued with the public interest. And I mean that in the largest and most favorable sense. Whenever you have a proceeding that also requires a certificate of public good, it means that at the end of the day you are very much involved in a public process."[8] Or as former CVPS President Robert Young said, "It's a commodity that is unique in its own right. In Vermont—and other places too but particularly in Vermont—it has this overlay of politics, of environment, of social good, of responsibility."[9]

Entergy got its first taste of Vermonters' deep interest and participation in electric energy issues as it moved towards finalizing the purchase of Vermont Yankee in the spring of 2002. That March, voters in seven towns passed resolutions rejecting Entergy's bid. The following year, the Brattleboro-based Nuclear Free Vermont circulated a second town meeting resolution—this time calling for the plant to close immediately and the electricity to be replaced with renewable sources. Entergy decided to contest this vote. As Vermont town meeting scholar Frank Bryan said, "A band of scruffy college students hanging from the fences of a nuclear power plant could be dismissed, but a town meeting of farmers and loggers and teachers and owners of mom-and-pop stores voting to disallow a nuclear plant in their village was something else indeed."[10]

Both sides worked hard at the town level to promote and oppose the resolution. Entergy spent heavily, investing more than $200,000. The company produced educational materials highlighting the plant's jobs and economic investment and charitable contributions. In the end the results were mixed enough that both sides claimed victory. The resolution passed in ten towns and failed in five (one town tied). But, because Entergy won the larger town of Brattleboro, overall votes favored keeping the plant open—2,849 to 2,579.[11]

Boosting the Plant's Power

Immediately following the purchase, Vermont Yankee's new owners moved forward on several fronts. The owners spent $8 million upgrading the plant's security, hiring a national firm, Wackenhut, to replace the Windham County Sheriff's Office.[12] They brought in teams of nuclear technicians from various plants to increase the plant's efficiency, building on their expertise and ownership of plants of similar design. And, shortly after taking ownership, they requested permission to increase the plant's output, filing an application with the Public Service Board in March 2003.

The request drew instant opposition from local citizens and the New England Coalition. Increasing the plant's energy output would only add stress to already-aging components, they argued. The Coalition's chief spokesperson and technical expert, the sharp-tongued Ray Shadis, told a reporter, "What they are really attempting to do here is to pack an additional nuclear reactor in the shell of an old aging reactor with aging components."[13]

Shadis, a loquacious, fierce and uncompromising advocate, had first engaged with the industry from his home near Maine Yankee on Maine's ocean coast. Shadis and the group Friends of the Coast succeeded in raising a number of safety issues regarding the 1972-vintage plant's operations, leading to an extensive review by the NRC. The NRC identified a long list of needed safety and maintenance improvements. Rather than pay for the improvements, the Maine utility owner voluntarily closed the plant.[14] Working part-time, Shadis led the technical side of the Coalition's work from his home in Maine, relying on his extensive self-education and personal motivation. (A constant presence in the news media's coverage of Vermont Yankee, Shadis is cited almost 500 times in the 1,400 stories analyzed here.)

Figure 7. The New England Coalition's Technical Advisor Ray Shadis holds a replica of a nuclear fuel rod at a public forum with the NRC in Brattleboro. Photo by Glenn Russell, *Burlington Free Press.*

Concern about the power increase also ran high with citizens living near the plant. More than 350 local residents attended a public meeting with Nuclear Regulatory Commission officials in June 2003 at the Vernon elementary school gymnasium, within shouting distance of the plant. The packed auditorium overflowed with citizens, and tempers ran hot. NRC officials told the crowd that evidence of the impacts of power boosts at older plants was limited, but that safety would not be compromised. The audience shouted their approval when one legislative opponent called it the equivalent of "putting a bigger engine in an old car."[15]

The NRC eventually signed off on the increase, rejecting the New England Coalition's arguments that the added stress would decrease safety margins. In the end, the NRC review took more than 11,000 staff-hours, the most extensive review of a power increase to date. The Coalition's Shadis blasted the results: "That is just a tribute to the bureaucracy; they are talking about a two-year paper chase, they are talking about 11,000 hours of sharpening pencils, 11,000 hours of going to the water cooler, 11,000 hours of putting on public relations events."[16]

The Coalition's primary concerns related to the additional stress that increasing power output would have on what the group argued were already-aging components. As Entergy slowly increased the plant's output, news stories focused on excessive "vibration" and "noise" causing Entergy to pause as the plant powered up. For the first time, terms such as aging and deteriorating had started to enter the public discourse about Vermont Yankee.

As Entergy proceeded with the uprate, the company requested state permission to store radioactive waste outside of the plant because the spent fuel pool inside the plant was reaching capacity. The 1977 Vermont law that advocates had fought for requires state approval of radioactive waste storage in Vermont for nuclear power plant owners other than the then-owners of Vermont Yankee. Entergy first argued that the legal right had passed from the plant to them, but lost that argument, turning then to the Legislature to amend the law. The subsequent law, Act 74, gave Entergy permission to store the waste in casks outside the plant and created the Clean Energy Development Fund to promote renewable energy projects. Entergy agreed to make payments totaling more than $25 million towards the fund goals of accelerating economically and environmentally sound electricity resources.

In return for permission to increase plant output and waste storage, legislators wanted added benefits for the state. As one legislator said during the debate: "It really bothers me that a private company is going to be able to have an uprate ... so why wouldn't we charge them for at least making money and planting again dangerous material on our soil." Legislators said the deal that eventually emerged was a win-win for the state. The Public Service Board agreed, approving the permit and noting the economic benefits. Legislators though, added one caveat: they required Entergy to return to the Legislature for permission to store any additional waste on-site after March 2012.[17]

As the plant increased its thermal output in January 2006, Entergy filed the 20-year license extension application with the NRC. In Montpelier, the state's capitol, legislators gathered for the 2006 legislative session. Vermont's part-time Legislature meets from January to May each year. The 180 legislators share a small professional staff of attorneys who help draft legislation and other staff who manage committees, find witnesses and keep track of the paperwork.

The State House is completely accessible most hours of each day legislators are in session. There are no guards at the door and no baggage screening, with most areas of the building open to the public. Lobbyists, citizens and legislators wander the hallways, meeting for lunch in the cafeteria or in the dozens of committee rooms. Committee rooms are small but the culture allows and encourages citizens and lobbyists to participate. It's not unusual for a legislator to turn to a member of the audience during a committee meeting to check a statement or ask for opinion. Audience members, while often registered lobbyists (state law requires registration), could range from Entergy representatives to environmental group members, other energy stakeholders or simply be interested citizens.

As the session started, about four blocks from the Capitol in a run-down two-story office building, staffers at the Vermont Public Interest Research Group (VPIRG) kicked around ideas for the 2006 legislative session. Newly hired "energy advocate" James Moore joined the brainstorming. The tall, dark-haired, confident

Figure 8. Kenneth Theobalds, vice president for government affairs with Entergy Nuclear (right), testifies about the Vermont Yankee nuclear power plant in a legislative committee meeting at the State House in Montpelier. Photo by Glenn Russell, *Burlington Free Press.*

Moore had a broad portfolio on energy issues for the consumer and environmental group, including closing Vermont Yankee. One idea the organizers came up with was to require that Entergy receive legislative approval, in addition to PSB approval, for their license extension.

Giving the Legislature a voice seemed to be a natural extension of Entergy's 2002 commitment, parallel to the 1975 law requiring legislative approval of new power plant licenses and in line with the recent debate around Act 74. Vermont Yankee opposition groups and their allies in southern Vermont and the Legislature framed the issue as democratic, simply giving legislators a voice in this important decision. As one supporter argued, this bill "allows the legislature, people elected by the State of Vermont" to be part of the decision on the plant's future. Legislators were also interested in ensuring that the state benefit from hosting the nuclear power plant: "Because if the people of Vermont are not going to benefit from a sufficient amount of power at a good enough price or a long enough contract—there's no reason to have the plant operate in our ... region."[18] The new law would give them a chance to be part of that conversation.

Entergy resisted the added level of approval, testifying in opposition, calling it "unnecessary" and pointing out the company was already committed to seeking Public Service Board approval. Legislative leadership strongly endorsed the concept and the law passed the House by a margin of 130-0 and the Senate 18-5. Vermont's Republican governor, a solid supporter of Vermont Yankee, signed the new law—known as Act 160—into law just a few weeks before the first NRC public hearings on the license permit would take place in Brattleboro.[19]

"This is an important issue for Vermont," Governor James Douglas (R-Middlebury) told a reporter. "I don't think it's at all unrealistic that elected officials should weigh in." Entergy put a positive public face on the new law as well, with plant spokesperson Robert Williams saying, "We commend the Legislature, especially the House Natural Resources (and Energy) Committee, for putting a lot of effort into drafting a bill that should serve the state well."[20]

Entergy's opponents hoped the bill would provide them with an avenue to force the plant to close. Since the NRC had approved all completed relicensing applications to date, opponents saw little chance of success there. As VPIRG's Moore said, "The legislature was an arena where we thought we had a fighting chance. The court of public opinion is available to us in ways that the judicial courts are not."[21]

While Act 160 did require that some future Legislature would have to vote on the relicensing decision, it did nothing to increase the chances of opponents winning that vote. To win a legislative vote would take public and political opposition from places outside of Windham County where the reactor is located. And Vermont Yankee was rarely a topic of conversation outside of the county's borders.

That would change in August 2007.

A Powerful Photo

Vermont Yankee today is about 33 percent efficient, meaning that close to two-thirds of the heat generated in the nuclear reaction is returned to the environment as evaporated or warmed water.

Depending on the operating mode, the time of the year and conditions set in the plant's permit, cooled water is discharged back into the river. The cooling tower system is comprised of two banks of 11 towers, each with a large fan, a series of water-holding tanks and pumps and heat exchangers. A pipe carries heated water from the reactor to the top of the cooling towers, where it is sprayed upon corrugated metal over which the fans operate. The assembly is supported by a wooden structure and the pipes by wooden beams. The cooled water is then returned to the river, at temperatures that can range from 2 to 5 degrees above river temperature.[22]

One hot afternoon—August 21, 2007—as the temperature climbed into the 80s, plant technicians heard rubbing sounds coming from one of the cooling tower fans. Concerned about the noise, plant staff decided to reduce plant power to investigate. That afternoon, before they could act, a rotting wooden support beam carrying the water pipe in one of the towers dropped four inches. When the beam fell, the pipe burst and water then exploded through a hole in the side of the tower, cascading down in a torrent, carrying debris with it.[23]

Plant technicians immediately reduced power as workers sought to understand and repair the damage. Following established protocols, Entergy set up an Operational Support Center and informed the NRC's resident engineer.[24] The NRC notified the state's congressional delegation. The cooling tower collapse was not considered a safety event because that particular tower was not essential to plant operation. Because of the "non-safety" categorization of the incident, local public officials were not immediately informed, in some cases learning about the incident first from a newspaper reporter.[25] One of the workers at the site snapped a few photos.

That evening, Hans Mertens, the chief engineer at Vermont's Department of Public Service—the state agency that oversees Vermont Yankee—emailed state regulators and emergency management officials. Mertens pointed out first that it was not a "safety related condition." He then explained the accident this way: "VY discovered sagging wood support structure in the cooling tower area … Presumably a water leak somewhere in the cooling towers is causing the structural elements to weaken and sag ...VY plans to investigate and isolate the leak."[26]

Early the next morning, Mertens added these details: "The structural support in one cell collapsed yesterday and VY reduced power as planned. Currently they are at 38% power output. NRC was notified. VY is assessing damage and will determine next steps. VY reiterates, this is not a safety related condition."[27] Mertens attached an Associated Press article, filed the night before, which quoted NRC spokesman Neil Sheehan as saying the problem was a "sagging" and "deformation" in the wood.[28]

A few hours later that morning, Entergy Manager of Government Affairs Brian Cosgrove emailed DPS Commissioner David O'Brien. Cosgrove, a tall, burly and frank public affairs officer, was well known by and on good terms with state regulators and business leaders. Cosgrove had previously worked for the plant's Vermont owners and prior to that had spent eight years as the director of Vermont's Republican Party, working from an office in Montpelier. In his email to Commissioner O'Brien, Cosgrove included a copy of the photo that showed a cascade of water coming from the structure, writing, "I would ask that you limit circulation"[29] (see *Figure 9*).

Later that day the anonymous photographer emailed the photo to the New England Coalition's Ray Shadis. Shadis sent the photo to VPIRG's James Moore. And Moore circulated the photo within the advocacy community and to the news media. Independently, Bob Audette, a reporter at the *Brattleboro Reformer,* received a high-resolution photo suitable for displaying on the paper's front page and started to make phone calls. Said Audette, "We were on top of it as soon as we got the picture—on the phone with Yankee, with NRC, state regulators. And eventually, they let everyone come in and look at it, but by then, the water had stopped flowing. The pictures were dramatic, but not as dramatic as that initial picture we had with water pouring out of it...Which basically went around the world." [30]

On Thursday morning, the *Brattleboro Reformer* carried the photo at the top of the front page with a bold banner headline, "VY cuts output after cooling failure."[31] The Associated Press circulated a story around the state, "Cooling tower problem forces Vt. Yankee to reduce power"—the first of four consecutive Vermont Yankee news

Figure 9. August 21, 2007, cooling tower collapse photo snapped by an anonymous worker at the plant. The photo was posted to Google Images and circulated to the media.

stories posted by the AP that week.[32] Through the efforts of VPIRG's Moore and other advocates, the cooling tower collapse photos buzzed around the Internet. By the end of the day, the DPS's top lawyer, Sarah Hofmann, had received six separate copies in her inbox.[33]

Vermont Public Radio's Ross Sneyd described the photo for a radio audience this way: "The pictures show a 50-foot-tall section of wall that's made of large plastic and fiberglass louvers. It's collapsed onto a fence and water gushes from a ruptured pipe high up in the structure."[34]

Policymakers at both the state and the NRC seemed to immediately grasp the significance of the photo. NRC spokesperson Sheehan said, "The photos are pretty dramatic. There's no question about that."[35] Peter Shumlin, the Windham County Democrat who had returned as Senate president in January, drew an analogy with the war in Iraq: "When I saw the photograph of what just transpired, my first thought was I was looking at a shot of a bomb that went off in Baghdad. I'm extremely concerned that we have to have a really objective outside evaluation of the safety of the plant."[36]

Later, another prominent legislator said, "When you published that picture, and you took the image of the nuclear power industry; what does that connote in people's minds? Physicists, chemists, engineers, white labs coats: precision. And what you saw was rotten wood and broken sewer pipes. That's exactly what the image brought up. That's what it was. It was rotten wood and broken sewer pipes. And if they couldn't get that right, what else was going on in that place?"[37]

In Washington, Senator Bernie Sanders had the photo enlarged and placed on an easel to show to NRC Chairman Dale Klein. Klein said to Sanders, "When you see that visible collapse ... it is a concern and you would question the maintenance and safety and so forth."[38] In a follow-up letter, Klein stressed the non-safety aspects of the incident: "Although the portion of the cooling tower that collapsed *has no safety related function* to protect the reactor, I can understand how the photographs of the collapsed cell may be unsettling to the residents living near the plant" (italics added).[39]

Entergy's Response: Two Days Too Late

Entergy's immediate response followed the detailed protocols long established within the company and under the oversight of the NRC. The company opened an incident report and started an internal investigation. The NRC also opened an investigation, ultimately finding that the incident was of "low safety significance." Treating the finding as a "non-cited violation," the NRC found that Entergy had not properly recognized the importance of performing hands-on inspections of wood beams in difficult-to-access areas, such as inside the cooling tower structures.[40]

While Entergy officials followed a deliberate, technical and rational process to identify the issues that led to the collapse and to take necessary corrective actions, they did this work without engaging with or participating in Vermont's social, economic and policy culture. Entergy left the media, environmental groups and even state policy leaders out of their decision-making and information-sharing process. Entergy engineers responded as they were trained to do, carefully and cautiously, consulting with experts at the NRC, at the plant and within Entergy's extended nuclear family. Information about the incident was immediately circulated to other U.S. nuclear power plant operators under the industry's "lessons learned" program, but not to citizens living near the plant.[41]

Meanwhile, the photo and follow-up news stories continued to dominate headlines, causing a sharp spike in plant-related news coverage (see *Figure 2*). News stories that week examined the costs to ratepayers ("Vermont Yankee problem to hurt taxpayers"), Vermont's congressional delegation's response ("Congressional delegation calls for investigation into Yankee mishap"), reports on the damage ("Nuclear damage is worse than reported") and stories about the causes ("Officials search for explanations in Vermont Yankee tower collapse"). The *Brattleboro Reformer* published multiple stories in the week following the incident, including, presciently, "Recent VY incident may have legislative impact."[42]

On Friday, two days after the tower collapse, Entergy invited a group of reporters and state and local officials to the plant. Scenes of heavy equipment and hard-hatted construction workers played across the evening news.[43] Entergy officials sounded both confident and accessible, vowing to get to the bottom of the issue. "We take this situation very seriously, and we will understand the cause," said Entergy Director of Nuclear Safety Assurance John Dreyfuss as he showed reporters around the site. Seeming aware of the dramatic effect of the images, Dreyfuss added, "It looks much worse than it actually is. Most of this damage is cosmetic, the key thing is the structural members and that's where we're focusing the investigation."[44]

A few days later the state's top regulator, DPS Commissioner David O'Brien, visited the plant and announced that the cooling tower collapse was an "isolated problem." Unfortunately for Entergy, as O'Brien left the plant, a faulty valve forced an automatic shutdown—or scram—spurring another round of internal and external investigations and an additional wave of media coverage.

The NRC found this incident also to be non-safety related in the investigation that followed.[45]

The reporters covering the story saw it differently. The Associated Press's lead Vermont Yankee reporter, David Gram, who had been writing about the plant for more than 20 years, raised the age issue. In an article headlined, "Some wonder: Is Vermont Yankee showing its age?" Gram wrote, "Two mishaps within 10 days at the Vermont Yankee nuclear power plant have given ammunition to its critics, prompted questions from regulators and have the plant's owners working to reassure the public about its safety." Gram would come back to this theme with a story two months later headlined, "Aging nuclear plant raises questions about life without it."[46]

The editorial writers at the *Brattleboro Reformer* were less sanguine: "The reality is that while Entergy wants to run an aging nuclear reactor at 120 percent of its generating capacity until at least 2032, Vermont Yankee is slowly but surely falling apart. Incidents like last week's cooling tower collapse do not increase public confidence in Vermont Yankee. Neither do emergency shutdowns. While the public was not in any danger in either mishap, who's to say that we won't be as lucky the next time?"[47]

While the photo and accompanying news coverage provided a powerful visual image of an "aging" and "deteriorating" nuclear power plant, the incident also illustrated Entergy's lack of transparency. As *Brattleboro Reformer* reporter Bob Audette said, "The cooling tower collapse was definitely of concern. It wasn't necessarily related to safety. But I think the biggest reason, the biggest off-shoot, the biggest problem for Entergy came in that they didn't tell anybody about it. And technically they didn't have to tell anybody about it because it was non-safety related. But there was this whole question about communication. You want to be honest and upfront with people."[48]

By the end of 2007, both the Associated Press and the *Reformer* named Vermont Yankee's "troubles" one of the top stories of the year. From the *Reformer*: "The photos of water gushing out over a pile of rotted wood and twisted metal became a symbol of what opponents had been saying for years—Vermont Yankee is not safe to run for an additional 20 years ... Lawmakers suddenly started paying attention to what was happening in Vernon. And the whole debate over the plant's future began to shift ..."[49]

Tower Collapse as a Symbol

In the years following the cooling tower collapse, the event became journalistic shorthand for problems at the plant, immediately calling up images of an aging and unsafe nuclear power plant. The Associated Press, for example, referred to the accident more than 100 times in the 120 Vermont Yankee stories that followed. WCAX-TV, Vermont's most-viewed local news channel, included references to the plant's cooling towers in one-third of all its Vermont Yankee news stories starting in August 2007. Vermont Public Radio, the state's most-listened-to news radio network, also turned to the collapse template, slipping it into about 20 percent of their stories after August 21.

Over time the collapse took on an explanatory power—providing a reason for the plant's public meltdown. The photo locked in images of an aging, unsafe and mismanaged nuclear power plant. For many Vermonters, the cooling tower collapse was the first photo they had seen of the plant. And despite Entergy and the NRC's claim that there were no safety implications, reporters, policy leaders and citizens repeatedly connected the collapse with safety in the following years.[50] For example, Matthew Wald of the *New York Times* put it this way in his story about the Senate's 2010 vote: Vermont "decided to discontinue the power plant's license due to safety issues. Those include a water cooling tower collapse ..."[51]

The cooling tower collapse also spurred additional media coverage, particularly in statewide media outlets that had previously seen Vermont Yankee—a two-hour drive from Burlington—as "not in our neighborhood." For example, news stories written about Vermont Yankee by *Burlington Free Press* reporters totaled about 10 from 2003 through 2006. In 2007, the number of stories increased sharply to 17, with most of those coming after the tower collapse. Sam Hemingway, the tall, white-haired veteran *Free Press* newsroom reporter who wrote many of those 2007 stories, put it this way: "The cooling tower was a powerful issue. That seemed to us to be a big deal. That made it a statewide story."[52] The same pattern was true for other Burlington-area media outlets, including WCAX and VPR (see *Figure 2*).

Increased media coverage can serve to increase public concern. The powerful negative imagery associated with nuclear weapons (mushroom clouds) and nuclear energy disasters (Three Mile Island/ Chernobyl/Fukushima) remains close to the surface of public consciousness.[53] At a NRC hearing in Brattleboro two months after the accident, speakers mentioned the tower collapse more than 15 times. As one citizen said in obvious frustration, "We don't know where to go when the cooling tower collapses when NRC says it's out of our hands. It's non-safety related. And any time the reactor fails and it's considered non-safety related, there is no one saying, 'Oh, yeah, we are the ones accountable.'"[54]

The collapse also caused a negative spike in public opinion towards the plant. In a poll conducted for WCAX that fall, 58 percent of the Vermonters surveyed said they were "very concerned/ concerned" about safety issues at the plant.[55]

Vermont Yankee Story Starts to Grow

The year 2007 marked the beginning of the shift in the prominence of the Vermont Yankee story, from being a closely followed local story to its arrival as a story of significance to statewide media. While unfolding events at the plant contributed to this change, so did the work of statewide environmental organizations, which seized on the collapse template to promote the story of an aging and mismanaged nuclear power plant. To win the legislative vote to close the plant, these groups knew that they would have to engage legislators and citizens outside of southern Vermont.[56] And to do this they needed issues that would resonate with Vermonters hundreds of miles from the plant.

Deb Katz, director of the Citizens Awareness Network, a group opposed to the plant since 1991, said, "We made the decision that being in the southern part of the state would not win us the vote. We just couldn't focus locally. And we had to make it real that there were issues. In fact, not safety issues. Issues about reliability. Issues about trustworthiness. Issues about systemic mismanagement that could have meaning to other people."[57]

The water tower collapse became a central part of this strategy. As VPIRG's Moore said, "The collapse was an incredibly powerful image because it showed Entergy wasn't being straight with Vermonters, saying it was a minor thing, nothing really happened. And lo and behold something collapsed to the ground. It was a very visual manifestation of what we were concerned about which was that it was too old and it was falling apart."[58]

These environmental groups developed materials and campaign communications building on the collapse template in the years following the accident.[59] In 2008, VPIRG's summer door-to-door canvass distributed a brochure featuring the "collapse" photo to more than 30,000 Vermonters in 111 towns. That fall VPIRG delivered 12,000 postcards to Vermont legislators calling for Vermont Yankee to close (see *Figure 10*).[60]

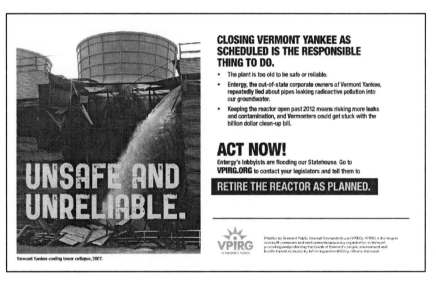

Figure 10. VPIRG postcard distributed to Vermonters during summer canvass operations in 2008. Courtesy of VPIRG.

When problems at a *second* Vermont Yankee cooling tower occurred in July 2008, these groups jumped on the story, further amplifying the "cooling tower collapse" template. In a press conference two days after this incident—also cited as non-safety by the NRC—the opposition groups reiterated their core narratives. VPIRG's director Paul Burns said, "It's not overstating things to say this plant is falling apart. It needs to be retired." Added the Citizens Awareness Network's Bob Stannard, "This plant is failing before our very eyes ... If these folks can't catch a water leak, heaven help us if they have a radiation leak at this plant." The incident, and later ones, also spurred a rebroadcasting of the original photo.[61]

The increased presence and advocacy of Citizens Awareness, VPIRG and other Montpelier-based groups such as the Toxics Action Center and Vermont League of Conservation Voters also signaled a change in the debate. While the New England Coalition focused on the regulatory process, these groups focused on grass-roots organizing, media outreach and legislative lobbying.[62] Instead of intervening in cases before the PSB and NRC, VPIRG, for example, aimed its professional organizers at winning the debate in the Legislature and in the media. From its base in Montpelier, VPIRG used its resources and skills to promote the "collapse" media template and broaden the story's impact.

This change can be seen in the news media coverage. Prior to 2007, reporters turned primarily to the southern Vermont-based New England Coalition to provide "balance" and challenger quotes for Entergy and state government officials in their news stories. Between 2003 and 2006, the Coalition received the vast majority of attributions cited to a citizens' group in opposition to Vermont Yankee. This changed in 2007, when the Coalition split attributions with VPIRG. And, in the three years leading up to the Senate vote, VPIRG was cited twice as often as the Coalition (see *Table 2*).

The change in emphasis and players can also be seen in the budgets and lobbying expenditures of the two organizations. Starting in 2006, after James Moore was hired, VPIRG resources allocated to opposing Vermont Yankee increased sharply, growing more than five-fold and peaking in the two-month period leading up to the State Senate's 2010 vote. In 2007, for example, VPIRG had revenues above $980,000, ending the year solidly in the black. That year the New England Coalition took in $139,000, ending the year with a

Table 2. Number of times New England Coalition or VPIRG or individuals associated with either group are mentioned in news media coverage in the *Burlington Free Press,* Associated Press, Vermont Public Radio and WCAX-TV between 2003 and March 1, 2010.

Organization	2003	2004	2005	2006	2007	2008	2009	2010
New England Coalition	87	173	83	63	75	65	20	24
Vermont Public Interest Research Group	4	0	3	1	57	103	59	45

Source: Media articles database. [63]

deficit of $23,500. Because of its focus on the regulatory process, the New England Coalition does not spend money on lobbying or legislative activities. VPIRG, on the other hand, reported spending $86,000 on lobbying in 2007, $133,000 in 2008 and $274,000 in 2009 and also in 2010. The Citizens Awareness Network also sharply stepped up its lobbying activities, hiring Bob Stannard as a full-time lobbyist and spending an average of $64,000 each year from 2008 through 2010. [64]

With a legislative vote seen as likely in 2008, Entergy also dialed up its legislative activities, increasing spending from $59,000 in 2007 to $361,000 in 2008, of which $244,000 was allocated to a paid media campaign.[65] The company added a second lobbying firm to its Montpelier contract lobbyist, expanding its presence in the Legislature. The company also hired additional consultants, including a specialist in press relations. And groups allied with the company also stepped up their advertising and promotion efforts, most notably the Vermont Energy Partnership. The pro-plant public relations campaign highlighted the plant's low-carbon, low-cost electricity. Full-page print ads in the *Burlington Free Press* and other newspapers read "Safe and Green: Safe, clean, reliable and economically sound nuclear energy is green power for the Green Mountain State. Let's keep it that way. Keep Vermont Green."[66]

Despite the spike in negative news coverage, most observers believed that the Legislature would approve relicensing Vermont

Yankee and that the Public Service Board would issue a permit to allow the plant to continue operating.[67] After all, the plant provided jobs to hundreds of Vermonters and circulated millions of dollars through the state economy in payroll, property and income taxes.

Notes

1 John J. Duffy, Samuel B. Hand and Ralph H. Orth, editors, *The Vermont Encyclopedia* (Burlington, VT: University of Vermont Press, 2003).

2 Grove Insight, "Building support for closing Vermont Yankee: Report of findings from a survey of 600 Vermont voters statewide with oversamples in Addison and Caledonia counties," January 19-22, 2010. Question wording: *"Generally speaking, do you support or oppose nuclear power?"*

3 Robert Putnam, *Bowling Alone*, (New York, NY: Simon & Schuster, 2000).

4 Duffy, Hand and Orth, *The Vermont Encyclopedia*.

5 Bryan, *Real Democracy*. Note: Norman Rockwell's famous painting, "Freedom of Speech," was based on a town meeting in Arlington, Vermont, where the Rockwell family lived from 1939 to 1953.

6 Ibid.

7 *The Vermont Encyclopedia; Real Democracy: The New England town meeting and how it works* (Chicago: University of Chicago Press, 2004); Frank Bryan and John McClaughry, *The Vermont Papers: Recreating democracy on a human scale* (White River Junction, VT: Chelsea Green Publishing, 1989); Cheryl Morse, in discussion with the author, January 13, 2012.

8 Steve Terry, in discussion with the author, November 10, 2003.

9 Robert Young, in discussion with the author, August 24, 2011.

10 Ibid.

11 Susan Keese, "Windham County towns split on Vermont Yankee resolution," Vermont Public Radio, March 5, 2003.

12 Justin Mason, "Officials give assurances of nuclear plant's safety," *Brattleboro Reformer*, January 5, 2004.

13 Susan Keese, "Public Service Board hears Vermont Yankee testimony," Vermont Public Radio, April 29, 2003.

14 Associated Press, "Former nuclear consultant: power boost violates rules," January 14, 2004; Stephanie Cooke, *In Mortal Hands: A cautionary history of the nuclear age* (New York: Bloomsbury, 2009); David Lochbaum, in discussion with the author, December 20, 2010. Note: Because the power boost required significant construction changes at the plant, state regulatory approval was required. The NRC would decide whether the power boost jeopardized safety.

15 Associated Press, "Some concerned by Vermont Yankee power expansion plan," June 12, 2003.

16 Christopher Graff, "Vermont Yankee gets green light to raise power," Associated Press, March 2, 2006.

17 *Entergy Nuclear Vermont Yankee, LLC et al v. Shumlin et al*, No. 1:11-cv-99 D. Vt (2012), http://www.vtd.uscourts.gov/Supporting%20Files/Cases/11cv99.pdf, pp. 14-22.

18 *Entergy Nuclear Vermont Yankee, LLC et al v. Shumlin et al*, p. 36.

19 Associated Press, "Douglas signs bill giving lawmakers say in nuke's future," May 21, 2006; Act 160 can be seen online here: http://www.leg.state.vt.us/docs/legdoc.cfm?URL=/docs/2006/acts/ACT160.htm and the vote totals for the Senate (bill S.124) are here: http://www.leg.state.vt.us/database/rclist/rclist2.cfm?Session=2006.

20 Associated Press, "Douglas signs bill giving lawmakers say in nuke's future."

21   James Moore, in discussion with the author, April 18, 2011.
22   "Entergy Vermont Yankee issues document," March 6, 2009; Vermont Agency of Natural Resources, "Amended discharge permit," Permit 3-1199, March 30, 2006, http://www.epa.gov/region1/npdes/permits/finalvt0000264permit.pdf.
23   David Gram, "Cooling tower problem forces Vt. Yankee to reduce power," Associated Press, August 21, 2007; Bob Audette, "VY cuts output after cooling failure," *Brattleboro Reformer*, August 22, 2007; Brian Cosgrove, in discussion with the author, January 4, 2012.
24   Hans Mertens, email, August 22, 2007; John Angil, email, August 21, 2007, at 3:44 p.m.; Neil Sheehan, email, July 13, 2011. Note: NRC has a resident engineer stationed at Vermont Yankee.
25   John Angil, email to state officials, August 21, 2007, at 3:44 p.m.
26   Hans Mertens, email, August 22, 2007, at 6:44 p.m.
27   Hans Mertens, email, August 23, 2007, at 7:44 a.m.
28   Gram, "Cooling tower problem forces Vt. Yankee to reduce power." Note: Sheehan appears frequently in the news articles examined here (664), following only Governor Douglas in the number of times cited by reporters.
29   Brian Cosgrove, email to David O'Brien, Aug 22, 2007, at 9:44 a.m.
30   Bob Audette, in discussion with the author, June 10, 2011.
31   Bob Audette, "VY cuts output after cooling failure," *Brattleboro Reformer*, August 22, 2007; Bob Audette, "VY probes tower failure," *Brattleboro Reformer*, August 23, 2007.
32   Gram, "Cooling tower problem forces Vt. Yankee to reduce power"; David Gram, "Vermont Yankee problem to hurt rate payers," Associated Press, August 22, 2007; Associated Press, "Congressional delegation calls investigation into Yankee mishap," August 23, 2007; John Curran, "Contract talks continue, strike possible at nuclear plant," Associated Press, August 24, 2007.
33   Sarah Hofmann, in discussion with the author, June 3, 2011.
34   Ross Sneyd, "Officials search for explanations in Vermont Yankee tower collapse," Vermont Public Radio, August 24, 2007.
35   Ibid.
36   Paul H. Heintz, "Recent VY incident may have legislative impact," *Brattleboro Reformer*, August 29, 2007.
37   Tony Klein, in discussion with the author, May 11, 2011.
38   Editorial, *Valley Advocate*, October 11, 2007. See photo posted at Philip Baruth's blog, http://vermontdailybriefing.com/?p=794.
39   Dale Klein to Senator Edward M. Kennedy, October 31, 2007, http://www.nrc.gov/reading-rm/doc-collections/congress-docs/correspondence/2007/kennedy-10-31-2007.pdf.
40   NRC, "Vermont Yankee Nuclear Power Station; maintenance effectiveness," September 30, 2007, http://pbadupws.nrc.gov/docs/ML0821/ML082100222.pdf.
41   Many scholars have written about the scientific approach to planning, see John Friedmann, *Planning in the Public Domain: From knowledge to action* (Princeton, NJ: Princeton University Press, 1987); Howell S. Baum, "Practicing planning theory in a political world," *Explorations in Planning Theory*, eds. Seymour Mandelbaum, Luigi Mazza, and Robert Burchell (New Brunswick, NJ: Center for Urban Policy Research, 1996), 365-382; Frank Fischer, *Citizens,*

Experts, and the Environment: The politics of local knowledge (London: Duke University Press, 2000). On utility planning, see James Throgmorton, *Planning as Persuasive Storytelling: The Rhetorical Construction of Chicago's Electric Future* (Chicago: University of Chicago Press, 1996); Aynsley Kellow, *Transforming Power* (Cambridge, UK: Cambridge University Press, 1996).

[42] Bob Audette, "VY tower collapse leads to calls for further safety review," *Brattleboro Reformer*, August 28, 2007; Audette, "VY probes tower failure"; Heintz, "Recent VY incident may have legislative impact."

[43] Kate Duffy, "Touring the damage at Vermont Yankee," WCAX, August 29, 2007; Bob Audette, "Investigation begins at VY," *Brattleboro Reformer*, August 25, 2007.

[44] Kristin Carlson, "State Official: Closer inspections needed at Yankee," WCAX, August 24, 2007.

[45] David Gram, "Nuclear plant sees automatic shutdown," Associated Press, August 30, 2007; Associated Press, "Public safety commissioner tours Vermont Yankee," August 30, 2007; NRC, "Vermont Yankee Nuclear Power Station; maintenance effectiveness."

[46] David Gram, "Some wonder: Is Vermont Yankee showing its age?" Associated Press, August 31, 2007; David Gram, "Aging nuclear plant raises questions about life without it," Associated Press, November 4, 2007.

[47] Editorial, "Who's minding the plant?" *Brattleboro Reformer*, September 1, 2007.

[48] Bob Audette, in discussion with the author, June 10, 2011.

[49] Randolph T. Holhut, "Year in review[1]," *Brattleboro Reformer*, December 29, 2007, http://www.reformer.com/ci_7831292.

[50] Although not directly related to safety, the NRC and others did cite the "cross-cutting" concept—where failures in non-safety components may have relationships to training, inspection and maintenance procedures that do relate to the safety features of the plant.

[51] Susie Steimle, "Who will decide the future of Vermont Yankee?" WCAX, March 11, 2011, http://www.wcax.com/story/14236529/who-will-decide-the-future-of-vermont-yankee; Matthew L. Wald, "Vermont Senate Votes to Close Nuclear Plant," *New York Times*, February 24, 2010, http://www.nytimes.com/2010/02/25/us/25nuke.html; Kristin Carlson, "Shut Down Vermont Yankee?" WCAX, August 19, 2008, http://www.wcax.com/story/8867500/shut-down-vermont-yankee.

[52] Sam Hemingway, in discussion with the author, July 6, 2011.

[53] Jon Palfreman, "A tale of two fears: Exploring media depictions of nuclear power and global warming," *Review of Policy Research* 23(1), pp. 23-43. For a scholarly analysis of the media template concept see Jenny Kitzinger, *Framing abuse: Media influence and public understanding of sexual violence against children* (London: Pluto Press, 2004).
Kristin Carlson, "State official: Closer inspections needed at Yankee."

[54] NRC, "Vermont Yankee Nuclear Power Station limited appearance: evening session," Docket No. 50-271-LR, October 11, 2007.

[55] Kristin Carlson, WCAX, November 13, 2011; RKS Research and Consulting, "GMP's Vermont Yankee relicensing survey," August 7, 2009.

[56] James Moore, in discussion with the author, April 18, 2011.

[57] Deb Katz, in discussion with the author, October 3, 2011.

[58] James Moore, in discussion with the author, April 18, 2011.

[59] Duane Peterson, in discussion with the author, April 6, 2011.

[60] Paul Burns, in discussion with the author, June 1, 2011.

[61] Darren Perron, "Vermont Yankee Fallout," WCAX, July 18, 2008, http://www.wcax.com/story/8700074/vermont-yankee-fallout; Kristin Carlson, "NRC releases Vt. Yankee cooling towers review," WCAX, October 14, 2008, http://www.wcax.com/story/9174213/nrc-releases-vt-yankee-cooling-towers-review.

[62] Paul Burns, in discussion with the author, June 1, 2011; Deb Katz, in discussion with the author, October 3, 2011; Ray Shadis, in discussion with the author, January 30, 2012.

[63] From the media database (1,409 articles). Includes all citations to organization and individuals employed by or introduced in conjunction with the organization. Note: Because the New England Coalition is based in the *Brattleboro Reformer's* coverage area, the Coalition continued to have a strong presence in *Reformer* stories throughout the case study. In this analysis we attempt to control for that local bias by not including the *Reformer*. The pattern is similar with the *Reformer* but not as marked, with the Coalition peaking at 287 mentions in 2007 and declining to 105 in 2010.

[64] The VPIRG spending is primarily paid staff who lobby as part of their policy issue work. The number is not broken out by campaign, so it includes staff lobbyists on consumer protection, good government, health care and other parts of VPIRG's public policy agenda. Vermont Secretary of State lobbyist disclosure forms, online at http://vermont-elections.org/elections1/lobbyist.html; James Moore and Paul Burns, in discussion with the author, April 18, 2011, and June 1, 2011, respectively; Department of the Treasury, Internal Revenue Service, "Return of organization exempt from income tax, form 990," http://www2.guidestar.org/organizations/03-0225975/new-england-coalition-nuclear-pollution.aspx#, http://www2.guidestar.org/organizations/03-0228267/vermont-public-interest-research-group.aspx.

[65] Vermont Secretary of State lobbyist disclosure forms, online at http://vermont-elections.org/elections1/lobbyist.html.

[66] Entergy Advertisement, *Burlington Free Press*, January 16, 2007, p. 3A.

[67] Tony Klein, Robert Dostis, Richard Smith and Paul Burns, in discussion with the author, May 11, 2011, May 27, 2011, May 9, 2011, and June 1, 2011, respectively; Kate Galbraith, "Climate enters debate over nuclear power," *New York Times*, May 30, 2008, http://www.nytimes.com/2008/05/30/business/30nuke.html.

FOUR

A Question of Value

Vermont Yankee's contribution to the state's economy is a powerful story, one Entergy highlighted in the regulatory and political processes around relicensing the plant. For almost 40 years the plant has provided one-third of Vermont's electricity, good-paying jobs and local and state tax revenues. As Entergy spokesperson Larry Smith said in 2009, "There is tremendous economic benefit for continued operation of Vermont Yankee. We feel we have a good story to tell the people of Vermont and the Legislature."[1]

Originally purchased for $180 million in 2002, the plant had an estimated value of $239 million by 2006.[2] By 2009, Entergy was running the plant 93 percent of the time, on average, generating between 4.7 million and 5.1 million megawatt-hours per year.[3] At prices of about $.04/kWh over the 10 years, the plant's gross sales averaged $200 million annually. Contracted sales with the plant's former owners accounted for most of the plant's output. Although Entergy was able to sell power from the uprate (approximately 100 megawatts) into wholesale markets, the remaining plant output was tied up in the 10-year contracts signed in 2002. The original sales agreement with Vermont utilities turned out to be a good deal for Vermont, as wholesale electric prices fluctuated between $.043 and $.088/kWh between 2002 and 2010.[4]

The plant's electricity business translates into major economic impacts in southern Vermont and to the state as a whole. In 2008, Entergy's payroll topped $55 million, the company paid state and local taxes of nearly $8 million and purchased more than $6 million of local goods and services.[5] Looking forward, a study commissioned by the Department of Public Service estimated that over the 20-year relicensing period, the plant could save Vermonters $2 billion in electricity costs, pay state and local taxes of $148.8 million and contribute an additional $7.6 million to local charities. Total potential benefits to the state were estimated at between $1.5 and $5.1 billion over the 20-year life of the license—a staggering amount in the context of Vermont's economic scale.

The Department's 2009 study laid out four areas where Vermont Yankee's continued operations would benefit the state: tax revenues, profit sharing after 2012, reduced power costs and the multiplier effect of these added dollars in the economy. Overall, the study's authors developed three potential futures: a high-impact case with benefits of $5.1 billion, a low-impact case ($1.5 billion) and a middle-range scenario ($3.6 billion) (nominal dollars). The analysis projected the middle-range scenario would produce about 1,208 jobs a year, of which almost half fell in Windham County.[6]

These are big numbers, representing a countable percent of Vermont's total GDP and a sizable contribution to state coffers. For example, each year the state was expected to collect more than $16 million in income, property and electricity taxes. However, the core of the expected savings to ratepayers came from the revenue-sharing agreement, negotiated when Entergy purchased the plant. Starting in 2012, Entergy had agreed to share half of the revenue above $.061/kWh (indexed for inflation) for 10 years after the plant was relicensed.[7] Based on the report's middle-range case, and making a number of assumptions about future power prices, the estimated average value to Vermont ratepayers was $93.9 million per year, for a 10-year total value of $938.8 million. The report also estimated reduced power costs through a long-term negotiated deal with Vermont electric utilities, looking at a range between 5 percent below average market prices, 15 percent and 25 percent. Taking the middle case, the authors estimated a net benefit of $178.2 million to ratepayers over 20 years.

The total benefits—$3.6 billion in the middle case—were emphasized by Entergy in the relicensing case before the Public

Service Board and in their advertising and lobbying efforts.[8]

When Entergy filed for their state permit in March 2008, their lead witness, Entergy Vice President Jay Thayer, summarized the economic arguments. Vermont Yankee "today supplies nearly one-third of the electricity consumed in Vermont at beneficial, below-market rates ... Since Entergy VY acquired and began to operate the VY station, the plant has operated at a Capacity Factor of 93%, providing power that is needed by Vermont and the region on a 24/7 basis ... Entergy VY is a major employer and economic force in Windham County and Vermont."[9]

Entergy and the company's allies strongly promoted these themes. Plant supporters frequently talked about the plant's reliability, critical contribution to Vermont's electricity mix, low prices, jobs, and income, payroll and property tax contributions. The company's biggest champion, Governor James Douglas, was particularly skilled at staying on this message. Arguing for relicensing the plant, Douglas said, "This is the cheapest power we have—we're talking about this in the context of the greatest economic crisis we've faced in some time and the key cost of doing business in Vermont and the rest of this country is the cost of electricity."[10]

Entergy's business allies, such as the Vermont Energy Partnership, also chimed in. "Because of Vermont Yankee's dependable and low-cost power, Vermonters have the lowest electric rates in the region, approximately 21 percent lower than elsewhere in New England," said Brad Ferland, the group's president.[11] In January 2009, the Vermont Energy Partnership released a report examining the "gloomy" consequences of closing Vermont Yankee, documenting the expected electricity cost and greenhouse gas emissions increases that would result.[12]

IBM, the state's largest private employer, was another key ally for Entergy. Keeping IBM happy had been very important to state politicians since IBM's Essex Junction plant first opened in 1957. With 5,500 well-paying jobs, IBM spends about $34 million a year on electricity to run its chip processing facilities. The company's lobbyist, John O'Kane, was a constant presence in Montpelier, reminding lawmakers of the importance of steady and low-cost electricity. "For IBM, the most important thing is that energy be reliable and affordable," O'Kane told a reporter in 2008. O'Kane was quick to remind lawmakers of his company's importance to Vermont: the annual payroll of IBM's Vermont workforce topped $250 million a year.[13]

Gearing up for a legislative vote in 2008, Entergy hammered home its economic progress message with a series of print and television ads. Advertising expenditures increased sharply, from $58,000 in 2007 to more than $260,000 in 2008. The "Vermonter to Vermonter" campaign ads featured Vermont Yankee employees talking about the plant: "Did you know that Vermont Yankee, its employees and their families bring millions in economic benefits to the state through purchases, taxes and payroll? Generous annual contributions to local charities. Nearly $5 million each year in education and general fund taxes. And hundreds of millions in energy costs savings. For more facts, visit safecleanreliable.com."[14]

Opponents Fight Back

Economic value was a powerful message and plant opponents knew they had to change the conversation. They had to reframe the debate. Instead of debating jobs and investments, opponents in citizens' groups and the Legislature attempted to turn the conversation to the story of an aging facility and an untrustworthy, out-of-state company. The cooling tower collapse had vividly illustrated a deteriorating nuclear reactor. Starting in 2008, Entergy gave opponents a hook to tie that story to their preferred narrative of an out-of-state company intent on making money on the aging nuclear reactor, putting Vermonters at risk.

A few months before Entergy submitted their license-extension application to the Vermont Public Service Board, the company announced plans to create a new subsidiary for its six for-profit nuclear reactors. Entergy initially dubbed the new company Spinco, but later changed the name to Enexus. The plants were mostly in the Northeast: Pilgrim in Massachusetts, FitzPatrick and Indian Point units 2 and 3 in New York, Vermont Yankee and Palisades in Michigan. Enexus would create an opportunity for investors to purchase stock in a company focused on for-profit nuclear power. There was much optimism from Entergy and Wall Street. In a press release, Entergy announced: "Spinco will be uniquely positioned as the only pure-play, emission free nuclear generating company in the United States."[15] Entergy's stock price rose 4.6 percent the next day.[16]

The extremely complicated petition submitted to the PSB spelled out 26 separate transactions involved in selling the assets, creating new entities and changing their responsibilities.[17] Essentially the proposal established a new company that would borrow money to purchase the six nuclear reactors from Entergy. For the new company to purchase the plants it would have to take on billions of dollars of debt, a concern to Vermonters. Instead of a company with the deep pockets of Entergy, the new owner's assets would be limited to the six reactors and more than $4.2 billion of debt. Legislators and opponents saw the proposal as a shell game, an effort to avoid the liability of paying for plant decommissioning by creating an underfunded and debt-ridden new entity, keeping the more valuable assets with Entergy.

The new highly leveraged company would be smaller and less diversified than Entergy, with fewer assets, less revenue and less income. Enexus, the new company, would rely for income on six nuclear reactors, all built in the 1970s, plants that could require substantial investments in the future. The debt-to-capital ratio would require that Enexus receive a non-investment grade credit rating. To sell the nuclear plants into the new company, Entergy needed approval of the NRC and regulators in Vermont and New York State.[18]

The second issue that emerged in 2008 had to do with the funds set aside to dismantle the plant and stabilize its future and long-term presence on the banks of the Connecticut River. At the time Entergy purchased Vermont Yankee, the Entergy subsidiary that owned the plant (Entergy Nuclear Vermont Yankee) agreed to fully fund plant decommissioning. Deconstructing and removing the radiation from a nuclear power plant is an extremely expensive proposition. For Vermont Yankee, the costs have been estimated at above $1 billion. When Entergy purchased the plant, they inherited a fund of $310 million into which the Vermont utility owners had been making regular payments. Entergy did not continue these payments. In the first years of Entergy's ownership the fund increased because of the increasing value of its holdings in the U.S. stock market. In 2008 and 2009, stock market declines wiped out most of those increases. At one point, Entergy was responsible for potentially $1 billion in costs, but had less than $500 million on hand.[19]

The increasingly skeptical Vermont Legislature conflated the two issues, the underfunded decommissioning fund and Entergy's efforts to spin off Vermont Yankee. "Now we're talking about this plant belonging to another unknown and perhaps less financially secure company," said State Senator Ann Cummings (D-Washington). Concerned that Vermonters would be "left holding the bag" for possibly hundreds of millions of dollars in decommissioning costs, Cummings's Senate Finance Committee drafted a bill to increase taxes on Entergy. Entergy's supporters called these new taxes unfair and said that opponents were changing the rules.[20]

Figure 11. Rep. Patricia O'Donnell, R-Vernon (standing), speaks against a bill to increase taxes on the Vermont Yankee nuclear power plant to fund the plant's decommissioning at the State House in Montpelier in 2009. Photo by Glenn Russell, *Burlington Free Press*.

These legislative battles came to a head in 2008 and again in 2009 with proposed legislation increasing taxes on Entergy to address the perceived shortfall. Legislation passed the House and Senate both years, only to be vetoed by Governor Douglas. The debate was acrimonious, with legislative leaders accusing Douglas of favoring the company over the interests of Vermonters and Douglas blasting legislators for attempting to drive up electric costs. Efforts to reach the two-thirds majority needed to override the vetoes fell short. Douglas stressed the plant's economic value in explaining his vetoes. In 2008, he wrote that the bill would "unnecessarily and substantially increase the future cost of electricity on both businesses and families." In 2009, he said the bill "threatens our economic recovery by unnecessarily increasing electric rates for consumers and businesses."[21]

As the debate between the Governor's Office and the Legislature over Vermont Yankee accelerated, key legislative leadership positions changed. In January 2007, Peter Shumlin (D-Windham) became Senate president, the top leadership position in the State Senate. Hailing from Putney, near where the plant is located, Shumlin knew the issue well. An effective political leader, Shumlin was quick to act and unafraid of conflict, fighting publicly and frequently with DPS Commissioner O'Brien and Governor Douglas around nuclear issues. When Shumlin talked about safety concerns, Douglas and O'Brien accused him of "scaremongering." "This kind of charade he's putting on and scaring the public is just outrageous," O'Brien said. Shumlin responded, "It's pretty unfortunate when the commissioner of public service becomes a wholly owned subsidiary of Entergy Louisiana. He fights for the raw deal being given Vermonters harder than Entergy."[22]

In the summer of 2008, State Representative Tony Klein (D-East Montpelier), a public opponent of nuclear power, became chairman of the House Energy and Natural Resources Committee — ground central for Vermont Yankee legislation throughout the decade. The previous chairman had been a quiet supporter of nuclear power, willing to negotiate with Entergy and producing several important compromise pieces of legislation, (e.g., the bill allowing dry-cask storage in exchange for $25 million to the state's Clean Energy

Development Fund).[23] Klein is a striking figure, short and wiry with piercing blue eyes and a hat often perched on his shaved head. Klein gave nuclear supporters and opponents a fair hearing, allowing all views to be represented, but made no secret of his own opposition: "I've hated nuclear power for a long time. Basically, what my parents taught me is that it was a technology created for destruction, for war—which it did quite well. And it's been looking for positive use ever since. And—economically and realistically—it just can't get there, because there is no answer for the waste."[24]

In this increasingly contentious environment, Entergy was now directly engaged in at least five time-consuming and demanding processes in various regulatory and legislative arenas. With the state Public Service Board, the company was pursuing the license extension and the complicated corporate restructuring. At the same time, Entergy was fighting with the Vermont Legislature on plant closure costs, the proposed spin-off and a legislatively required reliability audit.[25] And overlaid on all this was the license review with the Nuclear Regulatory Commission.

The Debate over the Deal

Back in 2002, Entergy and Vermont's two major utilities had signed a 10-year purchase and sales contract for about 250 megawatts of electricity. Starting at $.04/kWh, the deal lowered rates for most Vermonters, won praise from the state's leaders and put Vermont's electric utilities on track for the lowest electric rates in New England.[26]

Vermont utilities expected another good deal as they started negotiating with Entergy to purchase electricity beyond 2012.[27] The regulatory bargain between power plant owners and state regulators institutionalizes electricity as a public good. Since the early 1900s, state regulators have evaluated proposed electricity projects (e.g., plants, long-term contracts) to see if they meet broad societal goals. Under Vermont law, Public Service Board officials review the effect of a facility on reliability, aesthetics, the natural environment and the economic benefits before issuing a permit, known as a Certificate of Public Good (CPG). States' rights in these cases were reasserted in an important U.S. Supreme Court decision in 1983.[28]

Green Mountain Power and Central Vermont Public Service felt they had a strong bargaining position because the Board would require evidence of economic value, and legislators, who would also have a vote, would expect price concessions as well.[29] Early in the negotiations, Green Mountain Power CEO Mary Powell flew to New Orleans to meet with Entergy management. "So my feeling was, this is a no-brainer! I was in the camp that I felt like it would help them, it would help Vermont, it would help everybody if the value proposition could be powerful and clear as soon as possible."[30]

Figure 12. Green Mountain Power CEO Mary Powell with the company's plug-in hybrid electric car and solar array. Courtesy of Green Mountain Power.

Powell, who started working at Green Mountain Power in 1998, becoming CEO in 2008, could not understand why Entergy resisted the Vermont company's early efforts. Robert Young, the president and CEO at CVPS at the time, put it this way: "I don't think Entergy really took the negotiations seriously. And I don't know why. I think it was pretty clear we were making proposals to them that made a lot of sense. And they were rejecting those proposals. I think we had a fundamentally different view of what the risk/reward trade-offs were in the negotiations, and they had a very, very narrow view. And we just couldn't bridge that."[31]

A number of external factors may also have contributed to the inability to reach an agreement. The evolving market environment was at least partially to blame. Prices were falling as a result of the impending recession and the discoveries of vast quantities of natural gas close to New England, which were forecast to provide downward pressure in electric prices for years to come. By the time Entergy had crafted a deal, it was out of date with forecasts. Compounding this was Entergy's desire to fold in the revenue-sharing agreement, which was also dependent on market price forecasts.[32] At the same time, relationships in the state had begun to deteriorate. Looking back, Powell's predecessor at GMP, Chris Dutton, said: "Entergy I think, and this is kind of human nature, were feeling beleaguered, put upon by Vermont and it flopped over into their attitude and approach toward negotiating a new transaction."[33]

As the utilities and Entergy discussed a new power deal, state regulators started asking about the lack of an agreement in the relicensing case—questions that would spill over into the public arena. Entergy argued in the case that a new agreement for selling electricity to state utilities was not necessary for the PSB to issue the permit. Based on the DPS's report cited earlier, Entergy presented the annual benefits of the already-existing revenue-sharing agreement at $938 million. The company pointed out that the continued operation of Vermont Yankee would result in more than $2 billion in added income to the state, including $300 million in increased tax revenues and the annual employment of more than 620 workers.[34]

Entergy's lead witness in the case, Vice President Jay Thayer, wrote that "these benefits alone provide an adequate economic basis for the Board to issue a CPG [certificate of public good]. It is not necessary or reasonable for the Board to condition the CPG on the realization of additional benefits, and it would be risky to do so. Vermonters have a lot to lose if VY does not continue to operate."[35]

The contained and careful Thayer, with dark-framed glasses set on a square face, was the public figure of Entergy during this time, testifying before the PSB and in legislative committees. Recently appointed vice president for nuclear operations, the professional engineer had responsibility for the license extension at Vermont Yankee as well as the license-renewal processes for Entergy's midwest and northeast nuclear plants.[36] It was Thayer's job to shepherd the Vermont Yankee license applications through these multiple forums.

State Regulators Want A Good Deal for Vermont

Entergy had two sets of state regulators to convince. Ultimately they needed the approval of the judges on the Public Service Board. But prior to that, and critical to success with the Board, they needed to convince the regulators at the Department of Public Service. As the agency that represents ratepayers in these contested legal processes, the DPS's testimony is influential. Participating in these contested processes is expensive and difficult for most everyday citizens and organizations. For this reason, the Department plays a critical role representing ratepayers before the Board and billing back their expenses to utility parties or the state's electric consumers. The two agencies are located together on the top three floors of 112 State Street in Montpelier, down the street from the State House, with the larger DPS on the second and third floor and the Board on the top floor. The staff of the two agencies know each other well, seeing each other on the way to lunch, in the elevator or on the stairs, at noontime basketball and out running errands on the city's two main streets.

After winning election in 2002, Governor Douglas appointed David O'Brien to be DPS commissioner, and O'Brien served in that role through 2010 when Douglas left office. A former financial analyst and economic development officer, O'Brien strongly supported the plant "as long as it was safe."[37] O'Brien's ideological differences with legislative leaders and his aggressive style put him in disfavor in the Capitol building, where Democrats controlled both houses. Some legislators saw DPS as "friendly" to Entergy, and O'Brien and legislative leaders were frequently at odds over state policy towards the company.[38] Despite the caustic air between O'Brien and legislative leaders, in this case both were in agreement. Entergy had to offer the utilities, and ultimately the state's citizens, a good deal.

The point man on outlining the economic benefits of Vermont Yankee to the state was David Lamont, the DPS's director of regulated utility planning. After 25 years at the Department, Lamont was a popular employee, friendly and approachable, a passionate Red Sox fan with a long history in Vermont utility planning. Every single major utility case in Vermont in the last two decades had touched his desk — sometimes burying his telephone under inches of paper.

Much of the PSB process consists of filed testimony and rebuttal testimony. Utilities and the parties make their cases and question each other through these written materials. There is also oral testimony where lawyers and board members can cross-examine witnesses. In his initial position paper, Lamont argued that "the central equation" is that if the state is to host the nuclear power plant, Vermont ratepayers should be "afforded a materially favorable power supply agreement in return." And, Lamont said, without such an agreement, a new license for the plant was not in "the general good of the state."[39]

In response, Entergy argued that the resource-sharing agreement that they had agreed to in 2002, plus the payroll and property taxes and local charitable contributions and other factors, met the Board's criteria for public good. As the company wrote in its reply testimony, "With over one billion dollars of beneficial economic impact, the additional, likely substantial value of the RSC (resource sharing) and the VY Station's critical contribution to maintaining the nation's lowest electrical-carbon footprint, Continued Operation will provide a unique combination of benefits to Windham County and Vermont as a whole."[40]

The Department of Public Service, the advocate for Vermont consumers, remained unconvinced. The DPS's final brief in the summer of 2009 recognized the plant's substantial economic benefits but argued against issuing a license extension. Because of the lack of an agreement between Entergy and the state's utilities, DPS concluded that operating Vermont Yankee beyond 2012 would not promote the general good: "without a favorably priced PPA available to all Vermont utilities, the Board should decline to issue Petitioners the requested authority for an extended operations period for VY."[41] Board members also seemed skeptical of Entergy's claims.

Jim Volz, Vermont's Public Service Board chair, is a distinguished-looking jurist, with a trim silver-and-grey beard and a careful air. Like many of the characters in this story, Volz has participated in decades of the state's energy debates, leading the DPS's legal team under Republican and Democratic governors before being appointed PSB chairman by Governor Douglas. Civil, cautious and well-respected, Volz presided over the permit hearings. During oral testimony, Volz asked Entergy about the lack of a deal with the state's electric utilities: "What's the benefit to Vermont of buying power from you at market rates when we can buy power from the market at market rates?" Thayer responded that the agreement to split profits above $.061/kWh remained in effect. Volz then pointed out that since that agreement already existed, Entergy was offering nothing new. "That's correct," Thayer responded.[42]

Legislators and Public Start to Ask about the Deal

While Entergy's arguments within the case may or may not have persuaded the Public Service Board, the debate had started to spill over into the public discourse, undercutting Entergy's positive economic story. Lamont's February testimony was characterized this way by the Associated Press's David Gram: "The Vermont Yankee nuclear plant hasn't demonstrated it should get a 20-year license extension, a state agency said Wednesday, citing the absence of a deal that would enable the plant to sell power to Vermont utilities after 2012."[43]

Legislators also joined the discourse. Rep. Tony Klein told a reporter at the beginning of the 2009 legislative session, "Why should we bother to think about reliability and health and safety if the economics aren't there?"[44] Legislators used the lack of an agreement as a reason to postpone the legislative vote beyond

2009, a vote many observers think Entergy might have won had it been held then.[45] As Gram wrote, "Legislators want to know what price Vermont Yankee will be offering to sell electricity if the plant continues to operate before they vote on the license extension."[46]

In early 2009, Entergy's Thayer told Klein's committee that the company was days away from a deal. As Klein tells the story, Thayer "testified in my committee on a Friday, and we said to him: So when are you going to have a deal? Show me the money! What's the deal? How come you don't have a deal? Let's see what the deal is. And he stood in my committee, on record, and he said, I'm going down to Louisiana tonight and I am coming back here Monday with a number. I haven't seen or heard from him since. And I haven't seen or heard from any member of the corporation in my committee since."[47] Thayer's commitment to Klein's committee was covered by Gram, and the failure to reach an agreement was cited in more than 100 news stories that followed.[48]

Pressure for an agreement continued to grow in the summer of 2009. In a story headlined "Lawmakers to nuke plant: Get power deal by Nov. 1," the AP's John Curran wrote, "State lawmakers on Tuesday implored the Vermont Yankee nuclear plant to reach a new contract with its electric utility clients by Nov. 1, saying the Legislature needs to know what ratepayers will be charged before it acts on the plant's bid for a 20-year license extension." The story quotes House Speaker Shap Smith: "If they can't show … that there will be continued lower power prices then it's not clear why we would have the conversation about continuing to operate Vermont Yankee."[49]

Although the utilities and Entergy continued to talk, they never reached an agreement. Running out of time, but needing to put something before the Board, Thayer submitted a proposed agreement to the Board on December 19, 2009—about two weeks before the start of the legislative session that would decide Entergy's Vermont future. In his letter to the PSB, Thayer proposed selling 115 megawatts at a starting price of $.061/kWh.[50] Because of the recession in New England and the abundant supply of natural gas, wholesale electric prices were about $.05/kWh, meaning that utilities could already buy electricity from other sources for less than Entergy was proposing, at least in the short term.

To further complicate the company's case, Thayer's offer was made by Enexus Energy Corp, the proposed Entergy spin-off. Although Enexus had not yet been approved, Entergy conditioned its proposal on the PSB's certification of the restructuring plan. The offer was publicly ripped: "The kilowatt-hour price is too high, the proposal includes no provision for shoring up Vermont Yankee's depleted decommissioning fund and it was made by a company that in effect does not exist yet," said Rep. Tony Klein. GMP's Mary Powell added, "We see our job as getting a power contract that provides value beyond what is already in place. Today's offer does not add meaningful value to what we already have."[51]

Behind the scenes, the utilities and Entergy continued to talk. A new deal, much more favorable to Vermont utilities, was quite close, CVPS's Robert Young said later. A deal that would cement the jobs and economic messages and bring utility and business alliances to the forefront. Opponents continued to worry that with the backing of the governor, DPS and key business groups, this message would carry the day, winning both the legislative vote and the PSB's approval.[52]

Changing the Story Frame

VPIRG, the Citizens Awareness Network and other citizen and environmental groups continued to step up their organizing pressure in 2008, broadening the debate beyond southern Vermont and seeking to expand legislative opposition. In 2008, the groups banded together under a broad coalition called Safe Power Vermont. To win a vote to close the plant, opponents would need legislators from populous Chittenden County, home to one-quarter of the state's population, as well as votes from more conservative areas of the state north of Chittenden County. Engaging citizens and legislators hundreds of miles from the plant remained a challenge.[53]

Opponents increasingly started to brand the plant as managed by an out-of-state company, separating Vermont Yankee's jobs and economic messages from its corporate owner, Entergy. As one legislative opponent explained it, "We made a big differentiation between Entergy and Vermont Yankee. Vermont Yankee is the local people. And there was a great deal of respect for the employees and people who worked there. So we tried not to shoot the messengers. We tried to be very careful in talking about Entergy Louisiana. We were doing that, because of respect for the people that are here. But also psychologically, if you say Vermont Yankee in Vernon, there's this picture. But if you say Entergy Louisiana, I mean, "Oh my God, those sleazy people from Louisiana?"[54]

When Entergy initiated its 2008 "Vermonter to Vermonter" advertising campaign touting the plant's safety, environmental attributes, economic benefits and reliability record, VPIRG's director Paul Burns, a media-savvy organizer with 25 years of experience, immediately attacked. In a six-page letter to Vermont's attorney general, copied to all media outlets, Burns filed a consumer fraud complaint, arguing that the plant was not safe, clean or reliable. "The irony of a Louisiana-based corporation saying 'Vermonter to Vermonter' is indicative of the deception that is inherent in these ads," Burns said.[55]

Starting in 2008, language branding Entergy as "out-of-state" increased sharply, from 10 or fewer mentions between 2003 and 2007 to more than 50 times in 2010 (see *Figure 13*).[56] Senate President Peter Shumlin embraced this strategy, often combining Entergy with Louisiana or New Orleans when discussing the plant. "It's my judgment that the Douglas Administration and the Governor has lost their objectivity when it comes to *Entergy Louisiana*. Clearly, as evidenced by the vetoes, they will stand up for *Entergy Louisiana* stockholders at the expense of Vermont ratepayers and Vermont voters" (italics added).[57]

Evidence of the success of this approach can be seen in a survey of Vermonters in early 2010. The most unpopular group identified was "out-of-state utility executives"—not trusted by more than one-third (35 percent) of those polled.[59] Entergy's actions provided credibility for narratives underscoring the company's out-of-state status. Shumlin later would mention Louisiana 13 times in his speech on the Senate floor calling for the plant to close.[60]

Entergy's proposed spin-off of Vermont Yankee, the funding shortfalls in the plant closure accounts and continued operational

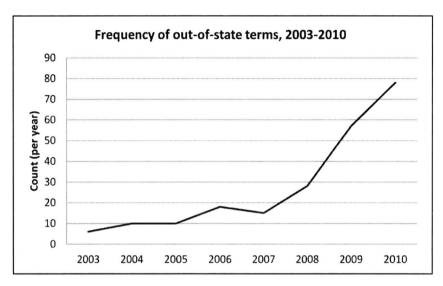

Figure 13. Framing Entergy as "out-of-state," 2003-2010 (n=132). Source, media database. [58]

issues at the plant contributed to this storyline. A senior Entergy vice president agreed, later writing the company's president that "opponents were successful in painting it [the spin-off] as a ploy by Entergy to shed decommissioning risk and ultimately stick Vermont taxpayers with the cost of decommissioning VY."[61]

Where are Entergy's Utility Allies?

As opponents moved to undercut the economic message and paint Entergy as an untrustworthy out-of-state company, the company's key Vermont allies, the electric utilities, sat on their hands. Vermont electric utilities are powerful forces in Vermont. The two electric utilities at the heart of our story, CVPS and GMP, are deeply immersed in the web of relationships that make up Vermont's policy culture. Together they employ more than 600 people, have direct relations with 250,000 Vermont customers, maintain a full-time staff of lobbyists in Montpelier, meet regularly with legislators and reporters and give money to local charities.[62] Here is CVPS's Young: "We are two companies very much embedded in Vermont. In every sense. In the communities we serve, in the political infrastructure that we have to deal with. In the regulatory infrastructure. And that is, in some respects, our lifeblood."[63] Or as GMP's Powell said, "We're Vermonters, we operate here, we know the customer base. We know the regulatory and political environment."[64]

In past nuclear power policy debates (for example, the power boost, dry-cask storage and plant purchase), the Vermont utilities were front and center, advocating for Entergy's position. But without a power contract, no utility or major business interest was "willing to capture the flag and just charge ahead for Entergy."[65] Entergy had no one to stand with them and help tell the story of economic value in the halls of the Legislature, in meetings with policy leaders, in day-to-day conversations.

Vermont utilities have skilled public relations teams and deep understanding of the local, political, social and regulatory environment. Entergy rarely asked them for help or advice and did not listen when they offered it.[66] A former legislator who went on to work at Green Mountain Power put it this way: "Unfortunately they had a poor PR campaign, but more importantly they had such poor relationships with people, they had no friends, and they kept losing friends. And because they were so secretive, not transparent, people didn't trust them."[67]

GMP's Powell said later that if they had a deal, the utilities would have been there: "I think it would have totally changed the dialogue because I think it would've been a whole group of us going in arm-in-arm saying, we get the concerns. We get the issues. But it's rated as being one of the top-performing plants in the country and this is a strong value proposition for Vermont."[68]

Entergy agreed. Again, evidence is provided by a memo from an Entergy vice president to the company president released several years later in the federal trial. In that memo, Entergy Vice President Curt Hébert wrote, "The lack of a completed PPA [power deal] with Vermont's two largest utilities and the generally negative way the utilities characterized the VY offer was a major obstacle to relicensing."[69]

Entergy had a great story, an existing plant in a strongly supportive community, hundreds of miles from state population centers. A plant that had been quietly operating for almost 40 years, providing high-paying jobs and investing millions in the state economy. But the story left out key characters; it left out the Vermont utilities and the Vermont regulators who believed that in return for hosting the plant, Vermonters should continue to receive a good deal for electricity. The economic value story got lost in Entergy's efforts to spin off Vermont Yankee, the debates around the decommissioning fund and ongoing operational and maintenance issues at the plant.

The plant's opponents, allied with legislators, were also able to knock Entergy off this message by reframing Vermont Yankee as Entergy-Louisiana, an untrustworthy company from "away." Ultimately, it was Entergy's actions that gave opponents a storyline to hang this narrative on, a storyline that would only grow as events continued to unfold.

Notes

1 John Curran, "VY critics renew campaign against relicense," Associated Press, November 19, 2009.

2 Kristi Ceccarossi, "VY assessment raises by 25% to $239M," *Brattleboro Reformer*, June 7, 2006.

3 GDS Associates, "Report to the Vermont Department of Public Service on the Vermont Yankee license renewal," Chapter 11, February 27, 2009, http://www.leg.state.vt.us/jfo/envy/7440%20Economic%20Cost%20Benefit%20Analysis.pdf, p. 26. Note: the plant produces less power in years with refueling outages.

4 US Energy Information Administration, "Wholesale market data," 2002-2011, http://www.eia.gov/electricity/wholesale/index.cfm.

5 "Entergy Vermont Yankee issues document," March 6, 2009.

6 GDS Associates, "Report to the Vermont Department of Public Service on the Vermont Yankee license renewal. Note: Study makes a number of assumptions, including the plant's operating capacity, the number of employees that live in Vermont and expected energy prices forward.

7 Vermont Public Service Board, "Memorandum of understanding among Entergy Nuclear Vermont Yankee, LLC, Vermont Yankee Nuclear Power Corporation, Central Vermont Public Service Corporation, Green Mountain Power Corporation, and the Vermont Department of Public Service," Docket no. 6545, March 4, 2002, http://www.leg.state.vt.us/jfo/envy/6545%20MOU.pdf.

8 GDS Associates, "Report to the Vermont Department of Public Service on the Vermont Yankee license renewal."

9 Vermont PSB, "Summary of prefiled testimony of Jay Thayer," Docket No. 7440, March 3, 2008, http://psb.vermont.gov/docketsandprojects/electric/7440/prefiled.

10 Kristin Carlson, "Vermont Yankee battle," WCAX, January 2, 2009, http://www.wcax.com/story/9609253/vermont-yankee-battle. Note: Douglas had multiple opportunities to make this case, as the most-frequently cited 886 times in the media coverage reviewed here.

11 David Gram, "VPIRG calls for closing Vermont Yankee in 2012," Associated Press, August 20, 2008.

12 Howard Axelrod, "An independent assessment of the environmental and economic impacts associated with the closing of the Vermont Yankee nuclear plant," prepared for Vermont Energy Partnership, November 17, 2008, http://www.vtep.org/documents/Axelrod%20Report.pdf; John Curran, "Consultant: Options limited for replacing Yankee," Associated Press, February 6, 2009.

13 Terri Hallenbeck, "Governor vetoes bill on Yankee," *Burlington Free Press*, May 8, 2008; Terri Hallenbeck, "IBM wants energy flexibility," *Burlington Free Press*, January 22, 2009.

14 Vermont Secretary of State lobbyist disclosure forms, online at http://vermont-elections.org/elections1/lobbyist.html; Vermont Yankee nuclear, "Smart," video posted on June 3, 2011, http://www.youtube.com/watch?v=EXc8mKL_6KU.

15 Entergy, "Entergy reports third quarter earnings and announces plan to spin off non-utility nuclear business," November 5, 2007, http://files.shareholder.com/downloads/ETR/1610695088x0x161590/80fa9e26-fdb9-4afc-9cea-2b4a42789073/ETR_News_2007_11_5_Earnings.pdf.

[16] Rebecca Smith, "Entergy nuclear spinoff taps rising plant values," *Wall Street Journal*, November 6, 2007, http://online.wsj.com/article/SB119432271390883672.html.

[17] Entergy petition to PSB, Docket no. 7404, January 28, 2008, http://psb.vermont.gov/sites/psb/files/docket/7404VT_Yankee_Reorg/Restructuring_Petition.pdf.

[18] PSB Order, Docket no. 7404, June 24, 2010. http://psb.vermont.gov/sites/psb/files/docket/7404VT_Yankee_Reorg/Restructuring_Petition.pdf.

[19] David Gram, "Decommissioning experts say estimates might be low," Associated Press, May 19, 2009; Fairewinds Associates, Inc., "Decommissioning the Vermont Yankee Nuclear Power Plant and storing its radioactive waste," prepared for Joint Fiscal Committee, January 12, 2011, http://fairewinds.com/content/decommissioning-vermont-yankee-nuclear-power-plant-and-storing-its-radioactive-waste.

[20] David Gram, "Vermont: Senate says decommissioning must be full before Vermont Yankee nuclear sale," Associated Press, March 21, 2008; David Gram, "Nuke decommissioning bill advances in House," Associated Press, March 27, 2009.

[21] David Gram, "Douglas administration questions Entergy reorganization," Associated Press, May 29, 2008; David Gram, "Vt. Gov. Douglas vetoes nuke decommissioning bill," Associated Press, May 22, 2009.

[22] David Gram, "Lawmakers want guarantees about nuke plant decommissioning," Associated Press, June 3, 2008.

[23] Robert Dostis, in discussion with the author, May 27, 2011; Tony Klein, in discussion with the author, May 11, 2011.

[24] Tony Klein, in discussion with the author, May 11, 2011.

[25] Associated Press, "Douglas administration hits nuke panelists," July 2, 2008.

[26] Vermont PSB, "Investigation into general order no. 45 notice filed by Vermont Yankee Nuclear Power Corporation re: proposed sale of Vermont Yankee Nuclear Power Station to Entergy Nuclear Vermont Yankee, LLC, and related transactions," Docket no. 6545, June 13, 2002, http://www.state.vt.us/psb/orders/2002/files/6545fnl.pdf.

[27] Mary Powell, Chris Dutton and Michael Dworkin, in discussion with author, July 5, 2011, June 3, 2011, and May 27, 2011, respectively.

[28] State of Vermont, "Act 248. New gas and electric purchases, investments and facilities; certificate of public good," Title 30, Chapter 5, http://www.leg.state.vt.us/statutes/fullsection.cfm?Title=30&Chapter=005&Section=00248; *Pacific Gas & Elec. Co. v. State Energy Resources Conservation and Development Commission*, 461 U.S. 190 (1983); Michael Dworkin, in discussion with the author, May 27, 2011.

[29] Ross Sneyd, "Senate bill would require legislative approval for nuke extension," Associated Press, March 14, 2006.

[30] Mary Powell, in discussion with the author, July 5, 2011.

[31] Robert Young, in discussion with the author, August 24, 2011.

[32] David Lamont, email, January 22, 2012.

[33] Chris Dutton, in discussion with the author, June 3, 2011.

[34] Vermont PSB, "Summary of prefiled testimony of Jay Thayer," Docket No. 7440, March 3, 2008, http://psb.vermont.gov/docketsandprojects/electric/7440/prefiled.

³⁵ Jay Thayer, letter to Susan M. Hudson, December 22, 2008, http://www.state. vt.us/psb/document/7440VT_Yankee_Relicensing/Supplemental/DPS_exh_ DL_1.pdf.

³⁶ Vermont PSB, "Summary of prefiled testimony of Jay Thayer."

³⁷ Rich Smith, in discussion with the author, May 9, 2011.

³⁸ Shay Totten, "The more things stay the same," *Seven Days,* January 21, 2009, http://www.7dvt.com/2009more-things-stay-same.

³⁹ Vermont PSB, "Prefiled direct testimony of David Lamont," February 11, 2009, http://www.atg.state.vt.us/assets/files/CLF%20Exhibit%2015B.pdf, pp. 21-24.

⁴⁰ Reply brief of Entergy Nuclear Vermont Yankee, Docket no. 7440, August 7, 2009.

⁴¹ Vermont DPS, "Brief of the Department of Public Service," Docket no. 7440, July 17, 2009, http://psb.vermont.gov/sites/psb/files/docket/7440VT_Yankee_ Relicensing/briefs/al_Brief_-_FINAL_Converted_to_Searchable_PDF.pdf.

⁴² Vermont Public Service Board, transcript of oral testimony, Docket 7440, p. 65.

⁴³ David Gram, "Vt agency argues against nuke license renewal," Associated Press, February 11, 2009.

⁴⁴ David Gram, "VY spokesman: Talks on future power deal still on," Associated Press, January 3, 2009.

⁴⁵ Tony Klein, Robert Dostis, and James Moore, in discussion with the author, May 11, 2011, May 27, 2011, and April 18, 2011, respectively.

⁴⁶ Associated Press, "Vermont Yankee decision delayed," December 7, 2008.

⁴⁷ David Gram, "Douglas hopeful on Vermont Yankee decision," Associated Press, February 27, 2009; Tony Klein, in discussion with the author, May 11, 2011.

⁴⁸ David Gram, "Vt. Senate chief questions Entergy spinoff," Associated Press, October 21, 2009.

⁴⁹ John Curran, "Lawmakers to nuke plant: Get power deal by Nov. 1," Associated Press, July 28, 2009.

⁵⁰ Synapse Energy Economics, Inc., "Preliminary analysis of Power Purchase Agreement from Entergy," prepared for Catherine Benham and Steve Klein, VT Joint Fiscal Office, January 15, 2010, http://www.leg.state.vt.us/jfo/envy/ Preliminary%20Analysis%20of%20Entergy%20PPA%20Proposal%2001-15-10.pdf.

⁵¹ John Curran, "Vermont Yankee files power price deal," Associated Press, December 18, 2009.

⁵² James Moore, in discussion with the author, April 18, 2011; Paul Burns, in discussion with the author, June 1, 2011.

⁵³ Ben Walsh, in discussion with the author, April 18, 2011; James Moore, in discussion with the author April 18, 2011.

⁵⁴ Jeanette White, in discussion with the author, June 11, 2011.

⁵⁵ David Gram, "Ad campaign prompts consumer fraud complaint," Associated Press, September 2, 2008.

⁵⁶ Media database, 1,409 news articles.

⁵⁷ John Dillon, "Shumlin calls for expert to improve Vermont Yankee oversight," Vermont Public Radio, June 19, 2009, http://www.vpr.net/news_detail/85205/.

⁵⁸ Analysis applied to 1409 news media articles. Words searched for included Missisipi, Louisiana, New Orleans, "out-of-state" and "non-Vermont."

59 Grove Insight, "Building support for closing Vermont Yankee: Report of findings from a survey of 600 Vermont voters statewide with oversamples in Addison and Caledonia counties," January 2010.

60 Center for Media and Democracy, "Under the dome: Vermont Senate vote on Vermont Yankee," February 24, 2010, http://www.cctv.org/watch-tv/programs/vermont-senate-vote-vermont-yankee.

61 Susan Smallheer, "Entergy ex urged 'open kimono,' according to testimony," *Rutland Herald*, September 13, 2011, http://rutlandherald.typepad.com/vermonttoday/2011/09/entergy-ex-urged-open-kimono-according-to-testimony.html.

62 CVPS, "Treating our communities as friends: 2010-11 citizenship report," http://www.cvps.com/AboutUs/10CitRep.pdf; Green Mountain Power, "Company facts and figures," http://greenmountainpower.com/about/commitment/2009-sustainability-report/facts-and-figures.html.

63 Robert Young, August 24, 2011.

64 Mary Powell, in discussion with the author, July 5, 2011.

65 Steve Terry, in discussion with the author, April 18, 2011.

66 Robert Young, Mary Powell, Robert Dostis and Steve Terry, in discussion with the author, August 24, 2011, July 5, 2011, May 27, 2011 and April 18, 2011, respectively. Note: For a review of utility participation in another deeply contested Vermont energy debate see Richard Watts, "Planning for Power: Citizen participation in the siting of a high-voltage transmission line in Vermont," (Doctoral dissertation, University of Vermont), February 2006.

67 Robert Dostis, in discussion with the author, May 27, 2011.

68 Mary Powell, July 5, 2011.

69 Curt Hébert, memo to J. Wayne Leonard, May 6, 2010.

FIVE

Erosion of Trust

A few days after New Year's, legislators returned to
Montpelier to launch the 2010 legislative session. The first
week is a busy and exhilarating time, as legislative leaders
meet with members, develop agendas and plot strategy. State
officials, citizens and lobbyists ply the hallways and the political
parties meet to hear their leaders' plans for the session. The key
event of the first week is the governor's State of the State address,
where the governor outlines state priorities. In Vermont, with a part-
time Legislature and limited staff, the governor with his control of
state agencies (the permanent government) plays a dominant role
in initiating public policy. The atmosphere was further charged by
Governor Douglas's announcement that he was stepping down,
marking an open race for the state's top job, the first time since 2002.

On a brisk winter day in early January, legislators convened in
the House Chamber for the State of the State speech. The House
Chamber is a spacious round room, directly below the Capitol's
dome. The chambers are dressed in colors of red and gold, with
historically accurate bright Axminister carpets. A gallery at floor
level stems from the main entrance doors. Citizens and lobbyists
can sit here close to the action, catching legislators as they pass
through the double doors out into the lobby. Against the walls near
the speaker's podium are rows of crimson-backed seats for members
of the State Senate, Vermont Supreme Court and other state officials
during joint sessions or important events.

The State of the State address is a much-watched event in

Figure 14. Gov. Jim Douglas receiving applause as he prepared to deliver his final State of the State address at the State House in Montpelier on Thursday, January 7, 2010. To Douglas's left is Lt. Governor Brian Dubie and House Speaker Shap Smith. Photo by Glenn Russell, *Burlington Free Press.*

Vermont. House members, senators and invited guests filled the Chamber as the governor was escorted in by the sergeant of arms and legislators.

This session was particularly poignant for Douglas, who had started his career 37 years earlier as a member of the House. During the following years he had served as representative, state treasurer, secretary of state and finally governor for eight years. Douglas had been on state ballots every two years for the last eight elections—a time period spanning 16 years. Reporters and onlookers crowded the balcony and visitor gallery that surround the House seats. Democrats vying to replace Douglas were present: Secretary of State Deb Markowitz, State Senator Doug Racine, State Senator Susan Bartlett and State Senate President Peter Shumlin. Looking on also was Lt. Governor Brian Dubie, Douglas's chosen successor.

In his last address, Douglas stayed with themes of fiscal

moderation and economic development, calling for balancing the budget without raising taxes and urging investments in areas that would create job growth. The state was facing a large $150 million deficit, and Douglas laid out several approaches to balance the budget without raising taxes: "A balanced and responsible budget is at the core of getting our state on track ... The single most consequential action we can take to encourage a healthy economy is to address the crushing weight of Vermont's tax burden."

About halfway through his speech, Douglas turned his attention to Vermont Yankee, calling on legislators to send the decision to the Public Service Board: "The decision about Vermont Yankee is central to our economic future and to maintaining a green energy portfolio. And it's a decision that should be left to the federal and state regulators—away from the political fray. For the hundreds of Vermonters employed at Vermont Yankee and many more who benefit from its economic impacts; for the thousands of Vermonters whose jobs depend on our competitive electric rates; and for a stable, clean energy future, this Legislature should vote to let the Public Service Board decide the case for relicensing."[1]

Listening to Douglas's speech, many political observers thought Vermont Yankee could still be relicensed. Entergy and the Vermont utilities could still broker a deal—a deal that would be compelling enough that it would trump everything else.[2] A poll conducted for Green Mountain Power in the summer had put public support for the plant at above 50 percent.[3] The Department of Public Service had signed off on the spin-off plan, as had the Nuclear Regulatory Commission.[4] The plant also had the support of Governor Douglas, DPS Commissioner O'Brien and several business groups. Entergy needed only a vote from the Legislature to allow the PSB to issue a decision. The company and its supporters believed that it was at the PSB that Entergy lawyers, resources and expert witnesses had a greater chance of success than in the political environment of the Legislature.[5]

But first Entergy needed an affirmative vote from the Legislature. Pressure was building on the legislators to send the decision to the Public Service Board—to let the professionals decide, to depoliticize the issue. As Entergy's Brian Cosgrove said,

"The Public Service Board process is designed to handle this kind of case, which is highly complex and politically charged."⁶ Still, the law was clear: the PSB could not issue a certificate of public good until the general assembly "grants approval."⁷

Calls for an Independent Review

Citizens, activists and some legislators had been calling for some time for an independent and comprehensive review of Vermont Yankee, a top-to-bottom examination of the plant's systems and components. Nuclear power safety issues are the responsibility of the NRC, and states are specifically preempted from reviewing safety issues under the 1954 Atomic Energy Act. Yet, citizens, environmental groups and some legislators did not trust the NRC's oversight.⁸ In Maine, complaints about the NRC's independence had led to an additional review, identifying enough problems that the plant's owners closed the facility.

While only the NRC can decide issues of safety, other issues such as reliability, economic impact and a plant's role in the state's energy mix can be reviewed by states. Legislators felt that one way to understand the advisability of continued operation of the plant was to examine the plant's reliability. If Vermont Yankee's systems and components were found to be experiencing frequent failures, or to be subject to failures in the future, it would shed some light on the plant's role in Vermont. Legislators and plant opponents who were concerned about the plant's safety also felt an independent review would provide important additional information. A plant that was not reliable might also be a plant that was not safe.

Legislators, distrustful of the Governor's Office, DPS and Entergy, were adamant that the review retain some independence. Governor Douglas and Entergy initially rejected the idea, stating concerns about the cost and duplication with existing state and federal review processes. As an Entergy spokesperson said, such a review would be "unnecessary and extremely wasteful in terms of time and money."⁹ However, the cooling tower collapses in 2007 and 2008 and other operational issues spurred Douglas to reconsider, setting the stage for a compromise with the Legislature. During the spring of 2008, as the governor fought with legislators over the decommissioning fund, the two sides negotiated a compromise.

That compromise culminated in Act 189, which was signed by Governor Douglas in June 2008 and required a "comprehensive vertical audit" of Vermont Yankee managed by the Department of Public Service but under the oversight of a newly created legislative panel. The audit was to take vertical slices of seven whole plant systems, looking at the systems from top to bottom to identify any reliability issues. Among the provisions, the law required the audit to examine whether the plant had an "underground piping system that carries radionuclides."[10] Leaks from underground piping systems carrying radioactive water had been found at more than one-quarter of the country's nuclear power plants.[11] Legislators were concerned that Vermont Yankee might experience similar leaks, potentially polluting area drinking water and the Connecticut River.[12]

Act 189 also created a public oversight panel to facilitate transparency, legislative oversight and communication. Governor Douglas, the House speaker and the Senate president were each to appoint one member and those members would then choose two additional members. House Speaker Gaye Symington chose Peter Bradford, a former NRC commissioner and a previous member of public utility commissions in both Maine and New York. Senate President Shumlin chose Arnie Gundersen, a nuclear engineer now living and working in Burlington. Douglas appointed Lawrence Hochreiter, a retired Westinghouse nuclear engineer. Together the three appointees added the Union of Concerned Scientists' David Lochbaum and Fred Sears of Pennsylvania State University. Hochreiter later died and was replaced with DPS's nuclear engineer William Sherman.

The antipathy between the Governor's Office, the DPS and the legislative leaders continued, despite the compromise that had led to Act 189. The hostility was probably enhanced by Symington's announced intention to run against Douglas for governor in the fall of 2008. The DPS's spokesperson, Steve Wark, blasted the Legislature's appointees: "These appointees today clearly have a bias against nuclear power."[13] The conflicts also extended to Entergy, which initially denied the panelists—all credible nuclear experts—

access to Vermont Yankee during a planned site visit, a decision that was later reversed.[14] The panel appointees actually worked quite well together over the next few years on a bipartisan basis. Sherman, for example, a thoughtful, thickly bearded and bespectacled engineer, had worked for more than 25 years at the DPS, which gave him credibility with legislators and Entergy officials. Sherman played an important role as the liaison between the DPS and the oversight panel.

Defining "Underground" Pipes

Soon after the passage of Act 189, DPS hired the Tennessee-based nuclear engineering firm Nuclear Safety Associates (NSA) to conduct the comprehensive reliability audit. Nuclear Safety Associates' nuclear-safety review staff had more than 300 combined years of experience working on NRC-regulated projects and was expert in meeting the demanding expectations of the NRC's technical staff.[15] Sherman and Nuclear Safety Associates developed a detailed scope of work that would guide the audit, reporting to the Legislature's oversight panel. Although Act 189 spelled out the broad direction of the audit, developing the detailed scope of work was a lengthy process which took place over the summer and fall of 2008. Sherman also involved Entergy in the discussions as the parties worked to decide which systems at the plant would be audited. A nuclear power plant is an extremely complicated industrial enterprise requiring expertise in chemistry, physics and electric and mechanical engineering, and auditors had a number of choices about which systems to investigate.

Legislators had specifically requested that the audit examine whether the plant had an "underground piping system that carries radionuclides."[16] Nuclear Safety Associates and the DPS, in consultation with Entergy, defined "underground pipes" to mean pipes buried in direct contact with soil and *not* pipes below ground but encased in concrete trenches. These interpretations led DPS nuclear engineer William Sherman to report to the Legislature's oversight panel that "there are no underground piping systems carrying radioactivity at Vermont Yankee." The technical experts at

Figure 15. A Vermont Yankee employee at the facility in Vernon. Photo by Glenn Russell, *Burlington Free Press.*

Nuclear Safety Associates turned first to reviewing a piping system below the surface of the ground, known as the condensate storage system. Entergy explained that those pipes were encased in concrete trenches, so the Nuclear Safety Associates staff turned instead to auditing the buried piping in the service water system.[17]

When released in late December 2008, the 415-page Nuclear Safety Associates report gave Vermont Yankee a favorable review, suggesting that the plant could operate for another 20 years. The auditors wrote that, "Overall, many station managerial and technical areas meet or exceed industry standards for performance. The station is operated and maintained in a reliable manner." Douglas and DPS Commissioner O'Brien touted the report's findings, stating that the "bottom line" was that the plant could run reliably for the next 20 years.[18]

The report specified that "there are no underground piping systems carrying radionuclides."[19] Despite this assurance, several independent observers remained unconvinced and kept asking questions. Chief among these was Arnie Gundersen, the tall, focused, white-haired nuclear engineer, now working both as the Legislature's special consultant and as a member of the public oversight panel. Gundersen's career in the nuclear industry traced back to the 1970s. Starting as an engineer, Gundersen rose to become a senior vice president with Nuclear Energy Services, managing and coordinating projects at more than 70 plants around the U.S. In 1990, Gundersen uncovered what he considered company malfeasance, "blew the whistle" and was subsequently blackballed and sued by the industry. The Gundersens moved to Burlington, where for about 10 years Gundersen worked as a high school math and physics teacher.[20]

Although supportive of nuclear power in general, Gundersen had been a persistent critic of Vermont Yankee because of its age, design and management. It was his knowledge and his understanding of the design features of the 1970s-era General Electric boiling water nuclear plants that initially drove him to ask about the "underground pipes." And it was this technical knowledge that was essential to legislative skeptics and plant opponents. In the increasingly contentious atmosphere between legislative leaders and the Governor's Office, legislators turned to Gundersen, former NRC commissioner Peter Bradford and others to provide independent assessments of DPS and Entergy reports. In July 2009, legislative leaders hired Gundersen and his partner, Maggie Gundersen, to review the Nuclear Safety Associates report and write an independent assessment.[21]

Gundersen initially accepted Entergy's statements that the plant had no underground pipes, thinking it unusual but not impossible. As he worked on the report, due in October 2009, Gundersen came across Entergy's 2008 Annual Effluent report. Entergy would only provide the 160-page report in hard copy, so Gundersen was unable to use the word search tools that he often employed to help sift through documents. Instead he had to read through the whole report. And in doing so, Gundersen came across a mention of underground pipes that had leaked in the past. Concerned that he had missed an important detail, in late July Gundersen emailed DPS nuclear engineer Uldis Vanags (who had replaced Sherman when he retired), pointing out that Entergy had told the Legislature's oversight panel

Figure 16. Members of the Senate Finance Committee listen as Arnie Gunderson (left) testifies about Vermont Yankee at the State House in Montpelier. Photo by Glenn Russell, *Burlington Free Press*.

that there "were no systems with underground piping that carry radioactivity at VY."[22] Gundersen wrote that he had since "become aware that there may indeed be underground pipes that do indeed carry radioactivity" at Vermont Yankee. Gundersen then asked Vanags "SPECIFIC question; *Is there underground piping that carries radioactivity at VY?*" (emphasis in original).[23]

Vanags passed the question on to Entergy and they consulted on how best to respond to Gundersen. Entergy expressed concerns to DPS about allowing Gundersen to reopen the reliability audit through the "back-door," pointing to the "potential chilling effect that the anticipated agenda-driven oversight (panel) would have on the willingness of people working at the site to raise concerns and identify issues."[24] There was no immediate response to Gundersen.

In the summer and fall of 2009, Entergy was engaged in multiple regulatory processes and staff were working nights and weekends to respond to information requests, review those requests with internal and external legal teams and pass them through their own internal vetting processes. Multiple staff at many different levels of the company were involved in responding to information requests. Both the NRC and PSB relicensing cases continued, which meant multiple rounds of discovery questions; company lawyers, technical staff and top managers were all actively engaged. Gundersen's information requests added to this workload. And the company had been skeptical of Gundersen's motives since he had first surfaced as a plant critic working as a consultant to the New England Coalition in 2003.[25]

Receiving no substantive response, Gundersen emailed Vanags again on August 7, asking about the "contaminated underground pipe issue." Vanags repeated the earlier responses from Entergy that they had no underground pipes. Still not satisfied, Gundersen on August 12 again asked Vanags: "I am aware of other underground pipes ... that are contaminated based on ENVY's own statements in published reports ... would you please ask ENVY to either elaborate on their previous statements that no such lines exist or to identify additional lines."[26] After another internal discussion, an Entergy manager responded, "Other than piping carrying gaseous material ... we have none." The gases had "very low amounts of contamination" and no way to contaminate groundwater. "Since this is not an item active in the review of CRA recommendations [the comprehensive reliability audit], we consider this issue closed."[27]

Still not satisfied with these answers, Gundersen's October report to the oversight panel pointed to the contradiction between Entergy's responses and its Annual Effluent Report. Gundersen spoke with Senate President Shumlin, who agreed to put the issue on the oversight panel's January agenda.[28]

Underground Piping Questions Continue

At the same time as Gundersen was pursuing Entergy on the question of underground pipes, state regulators and advocates were asking the same questions in the license extension case. The first day of oral testimony in an important PSB case is always a big event, bringing news reporters, interested observers and a slew of expert witnesses, state technical staff, board members and lawyers to Montpelier. Such was the case in May 2009 with the opening of oral testimony in Docket 7440, the Vermont Yankee relicensing case.

It was a mild spring day as the lawyers, expert witnesses and their clients filed into the hearing room on the third floor of 112 State Street, across the street from the Governor's Office and down the block from the State Capitol. As lawyers, witnesses and onlookers crowded in, PSB staff added another row of chairs to the cramped room. At the front of the room was a raised dais where the three Public Service Board members and their staff sat next to an American flag. To their right was the witness stand and immediately in front of the dais was a desk for the court stenographer. In front of the Board were a series of tables joined together in a U-shape, where lawyers representing the state, Entergy and other interveners sat. Behind the lawyers, arrayed around the room, were supporting staff, lead witnesses, expert consultants and managers. Also in attendance that first day were the lawyers representing the advocates, the Conservation Law Foundation's Sandra Levine, Jared Margolis with the New England Coalition, and staff and lawyers from the Agency of Natural Resources, Green Mountain Power, CVPS and the Windham Regional Commission. A handful of reporters, including the Associated Press's David Gram and Vermont Public Radio's John Dillon, and other interested observers filled the room.

Entergy had signed up Vermont's largest law firm as external counsel. Downs, Rachlin & Martin's John Marshall served as lead counsel and the firm's extensive resources were at Entergy's disposal. An experienced and able lawyer, Marshall had argued dozens of utility cases before the PSB in his 30-year legal career. In 2010, *Public Utilities Fortnightly* magazine named him one of 34 "Groundbreaking Lawyers" from across the nation in the field of energy law.[29]

Even for those who had been there many times, there was a tension in the air, a sense of anticipation. Although the prefiled testimony had been circulated, nothing beats a face-to-face encounter, with potentially contentious cross-examination, witnesses thinking on their feet and the possibility of unscripted answers.[30] These cases can be expensive (above $10 million for the applicant), time consuming and highly technical. Stakes were high: the Board's decision would decide Entergy's future in Vermont.[31]

Entergy's lead witness, Jay Thayer, testified on May 20, starting with the traditional oath to tell the truth. The room was silent except for the tapping of the stenographer, whispers from onlookers and people shifting in the blue-cushioned, curved metal chairs. DPS lawyer John Cotter started the cross-examination (Thayer's prefiled testimony was already in the record) by asking about the power purchase agreement (see Chapter 3). Cotter then turned to the standards for radiation levels at a decommissioned plant. He then paused, rustling through his notes. Board member John Burke asked, "Mr. Cotter, are you moving on, or are you still on the same line here?" Cotter responded, "Same line generally." He then asked Thayer, "Does Vermont Yankee have any underground piping that carries radionuclides?" Burke peered over his glasses from his seat on the dais.

There was a long pause—12 seconds—before Thayer responded: "The reason I hesitate is I don't believe there is active piping in service today carrying radionuclides underground. There was a line that was contaminated, radioactive liquid, which did leak back in the period before we purchased the plant, that line was abandoned. That is the reason for some of the contaminated soil on site. But I don't—I can do some research on that and get back to you, but I don't believe there are active piping systems underground containing contaminated fluids today."[32]

After 40 years in the nuclear industry, Jay Thayer had risen to become vice president of nuclear operations for the second-largest nuclear power plant operator in the U.S., a position that tasked him with multiple Entergy relicensing initiatives, including every step of the Vermont Yankee permit process. In the charged environment that would come eight months later, this answer would cost Thayer his job.[33]

Marshall suggested that the Board also ask the plant's top Vermont official, Site Vice President Mike Colomb, the same question the following week. And they did, asking, "Do you know if there's any underground piping at Vermont Yankee carrying radionuclides?" Colomb provided a similar response, focusing on the abandoned line: "I believe we had identified one pipe that was underneath the Chemistry laboratory that end—I believe leaked in the past, did contaminate some soil under the buildings, has since been sealed ..." Colomb also would pay a penalty for this answer, as he was later placed on administrative leave and fined by the company.

At the time, no one knew or understood the significance these answers would have. Yet, the advocates kept asking. A VPIRG lawyer submitted a question during discovery: "Have there been underground piping systems carrying radionuclides at ENVY at any time in the past?" Entergy responded: "There have not been any underground piping systems carrying radionuclides during ENVY's ownership of VY station."[34]

Entergy on the Defensive

On the same day that Governor Douglas called on legislators to send the license decision to the PSB, Entergy announced that a radioactive substance, tritium, had been discovered in one of the plant's monitoring wells.

The announcement of a potential radiation leak into area groundwater flashed across headlines in Vermont's newspapers and on the television news. Entergy and state regulators pointed out that the small amounts found presented no danger to the public. The state's radiological health chief, Dr. William Irwin, said that the radioactivity in the monitoring well should not be a concern because it was below the U.S. EPA's safety limits for drinking water of 20,000 picocuries per liter. NRC's Sheehan added similar statements. "Twenty thousand might sound like a lot, but it's not that high," Sheehan told a reporter from the *Brattleboro Reformer*. "You would have to drink some fairly large quantities every day." Gundersen though, called it a "big deal" because it suggested that a pipe or tank at the facility was leaking.[35]

In the following days, the story continued to grow as radio-activity levels increased and Entergy scrambled to find the source. A few days later, officials announced that tritium had been found in a second monitoring well. By the end of January, the NRC was reporting that the monitoring well had registered 70,500 picocuries per liter, more than three times the federal safety standard of 20,000 picocuries per liter. Still, the NRC, Vermont Department of Health and Entergy affirmed that there was no threat to human safety and no evidence that the tritiated water had migrated to household wells or the Connecticut River.[36]

Tritium is only harmful if ingested through food or water and has a fairly short biological half-life of 7 to 14 days—that is, half of the tritium would be excreted from the body within about 10 days.[37] Although tritium is a gas, it binds with the hydrogen atoms in water to form tritiated water. Tritium leaks have been found at more than half the nation's nuclear plants. The problem is frequent, well documented and a focus of both the NRC and the industry. The largest safety concern is that the tritiated water can leach into and contaminate area drinking water sources. It is for just this reason that Vermont Yankee has monitoring wells on the grounds near the plant.[38]

While the tritiated water did not ever appear to seep into drinking water, Entergy's inability to locate and close the leaks kept the story alive. In a 10-day span, dozens of news articles, television and radio stories covered different aspects of the leaking radioactive substance. Furthermore, while Entergy, the NRC and state health officials tried to quell public fears, frequent mentions of radiation and groundwater only increased residents' concerns. Daily stories about the radiated water, the spread of the water and the inability of Entergy officials to find the leak only added to public fears, triggering images of radiation poisoning.[39]

News stories placed the tritium leaks at Vermont Yankee in the context of other leaks at plants across the country. The AP's David Gram submitted a story that circulated across the U.S., headlined "Vt nuke plant leaks renew debate over aging plants." Gram wrote: "Radioactive tritium, a carcinogen discovered in potentially dangerous levels in groundwater at the Vermont Yankee nuclear plant, has now tainted at least 27 of the nation's 104 nuclear reactors raising concerns about how it is escaping from the aging nuclear plants."[40]

Federal officials and regulators in other states also started to pay increased attention to the unfolding situation in Vermont. Congressman Paul Hodes (D-New Hampshire) asked that his state, across the Connecticut River from Vermont, have added oversight to protect the health and safety of local residents.[41] And in New York, regulators concerned about events in Vermont started to ask additional questions about Entergy's proposed spin-off of the four reactors the company owned in that state. The New York officials were particularly concerned about the potential for similar leaks at these plants, and the potential inability of the new corporation to pay for needed fixes.[42]

Underground Pipes

Less than a week after the first announcement, the story took a sudden turn when Entergy confirmed that underground piping was a possible source—pipes that Entergy had previously said did not exist. Reporters, legislators and advocates scrambled to pinpoint the various times Entergy had discussed the existence of underground pipes. Ultimately, on nine separate occasions, twice under oath, Entergy had either said there were no "underground pipes" or missed an opportunity to correct the record.[43] Added to the stories about the search for the leak and stories about growing pools of radiated water, now came a series of news stories about the misleading statements. As a legislative champion of the plant said, "The fact that they admitted they misled people, yeah, that's not going to help." Commissioner O'Brien added, "I don't think we're even in trust-but-verify mode any more. I think we're just in verify mode now."[44]

Governor Douglas, while remaining a supporter of the plant, joined the criticism, saying, "I'm not sure they get it. I'm not sure they understand the great concern that Vermonters have, those of us in public office, and the people of our state all around Vermont about the situation at Yankee. Both with respect to tritium leak but even more importantly the breach of trust that's occurred between the company and our state."[45] Douglas asked for a change of leadership at the plant.

At a hearing on Vermont Yankee's license extension, PSB Chairman Jim Volz said, "It should go without saying, but perhaps Entergy needs to hear it anyway: Such conduct is absolutely unacceptable. Our decision in this proceeding, the cases presented by other parties, and the Legislature's own, parallel determination, all depend on Entergy

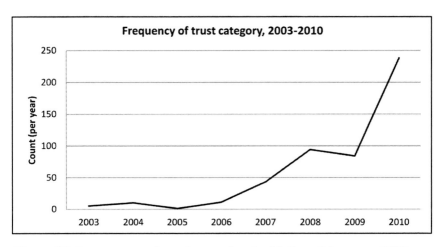

Figure 17. Occurrences of words associated with "trust" between 2003-2010 (n=217).[50]

providing timely, accurate, and complete information."[46] For Volz, this was strong language, particularly as he was the presiding judge in a still-open case on Entergy's license extension.

Entergy's Thayer issued a public apology for "failing to provide full and complete information about the presence of underground pipes at Vermont Yankee."[47] A few days later, Entergy placed Thayer on "permanent leave" and announced they had hired an outside law firm to investigate the underground piping statements.[48] On January 21, the firm placed a "litigation hold" on 65,000 documents within Entergy, freezing the files of 19 high-level employees, including Colomb and Thayer. Investigators started the painful process of interviewing each one of those employees, often with two lawyers present. A number of other investigations swarmed the plant. At the end of January, Vermont Attorney General Bill Sorrell announced a criminal investigation into Entergy's various statements.

Distrust Grows

Trust, meanwhile, had become the word of the day. As Douglas said, "What has happened at Vermont Yankee is a breach of trust that cannot be tolerated."[49] Words associated with trust increased sharply in news stories about the plant in 2010, appearing in more than half of the 185 articles published in the year. In the previous 820 stories, "trust" appeared in less than 10 percent of the articles (see *Figure 17*).

At the same time, the public accountability frame—a phrase or combination of phrases linking concerns about company oversight with company actions—also increased sharply, from about 10 to 15 percent of opposition frames to almost 50 percent. A frame in this case defines a problem (e.g., *Entergy is not accountable*) and suggests a solution (e.g., *close the plant in 2012*). Reporters, who also perhaps had grown skeptical of Entergy, adopted this frame without always attributing it to a news source. For example, by 2010 displays of this frame had doubled in Associated Press articles.[51]

By the end of January, Governor Douglas called for a "time out," asking the Legislature to postpone the vote given the "dark clouds of doubt" that should be cleared before proceeding.[52] As an Entergy government affairs official later summarized the situation, "high-profile news stories were running daily featuring accusations by plant officials that VY officials had intentionally given incomplete or misleading testimony." These news stories had a "corrosive effect" on even Vermont Yankee's strongest supporters, including Governor Douglas.[53]

Shumlin Schedules a Vote

On February 16, Senate President Shumlin scheduled the Senate vote on the plant's future for the following week. The next day, Entergy released the results of the company's internal investigation. The extensive investigation—an astonishing 2,300 man-hours expended in less than 30 days—found that while Entergy's statements could be seen as incomplete and misleading, they were not purposely deceptive.

With Thayer on permanent leave and other senior managers under investigation, Entergy turned over day-to-day direction of their public relations efforts to Curt Hébert, a senior vice president from the company's headquarters in New Orleans. Young, smart and articulate, Hébert was a rising star in the utility world. Appointed in 1997 to a seat on the Federal Energy Regulatory Commission, at age 38 he had been elevated by President George Bush to become the

Figure 18. Entergy Executive Vice-president Curtis Hebert Jr. speaks at a crowded press conference in the Cedar Creek Room at the Vermont Statehouse in Montpelier. Photo by Glenn Russell, *Burlington Free Press.*

commission's youngest chairman in history. Hébert left that position to become vice president of corporate relations for Entergy in 2001.[54] In the last few weeks before the Senate vote, Hébert had become the public face for the company, meeting with legislators and the press and directing the company's public relations strategy. It was now his job to deliver the results of the investigation.

Reading from a statement in a crowded press conference in the Capitol's Cedar Creek Room, Hébert said that although there was "no intention to mislead," there were "failures" in communication.[55] Hébert explained that Entergy, Nuclear Safety Associates and the DPS operated under a "shared understanding" of the meaning of "underground piping system" that differed from how the term might be perceived by others. Because the Entergy employees did not specify the context that they were operating in, Hébert said, their responses were "incomplete and misleading." Hébert announced that the company was placing five senior Vermont Yankee employees on administrative leave and reprimanding an additional six managers.

The law firm's extensive internal investigation detailed a distinction between underground pipes and pipes below the surface of the ground. Because of the shared understanding of the meaning of "underground" pipes, Entergy's answers were truthful, Hébert said. Entergy staff were answering questions in that context, interpreting "underground" to mean "only pipes that touched soil, not those encased in concrete." When Entergy employees were asked about "underground pipes" they did not include pipes below the surface of the ground if those pipes were contained within concrete trenches. This "shared understanding" evolved through discussions between Entergy, the Department and the outside auditor.[56]

The struggle to explain the difference between underground and under the surface of the ground swamped Entergy and their allies. The distinction was completely lost on most participants and observers. As lobbyist Stannard said, "You lie to someone in that building, you're done."[57] The company was further hampered by the pending criminal investigation and its own continued inability to find the source of the leak. "Nobody knew what else was under there or what more would come out," said a utility executive closely watching the scene at the time.[58]

Furthermore, although Hébert announced the report's findings and the punishment of 11 plant employees ("friends of ours") at the press conference, the report itself was not made public. Already under intense scrutiny for misleading statements, Entergy was immediately attacked for the lack of transparency. In a contrast with common practice at press conferences in Vermont, Hébert did not take questions. Opponents pointed out that the law firm hired for the investigation, Morgan, Lewis & Brockius, had strong ties to the nuclear industry, including previous work for Entergy. Entergy turned the report over to the Attorney General's Office, but reporters and legislators were skeptical. Since Entergy funded the investigation, chose the investigators and limited the investigation to Entergy employees, the results were seen as suspect.[59]

Two months later the report did become public. The 124 pages provided a fascinating snapshot of a company feeling besieged and buffeted by multiple information requests, struggling to respond quickly, properly vet answers and keep various key parts of the operation informed. In this pressure-cooker environment, company officials made a number of mistakes, highlighted in the 18 months of documents examined by the legal investigators. Senior staff were asked to approve detailed responses to discovery questions overnight, ("another Vermont problem," read the message line in one email). Managers admitted to signing off on testimony they never fully read. Flags raised about faulty testimony and legal responses were ignored. One senior manager, recovering from an injury, worked at home for six weeks.[60]

After Thayer testified to the Board that there were no active underground pipes carrying radioactive fluids, company staff recognized a contradiction between Thayer's statement and the actual facts on the ground at the plant. A similar "inconsistency" was noted between company piping systems and the answer provided to VPIRG. Emails flew around the facility, but the right people were not always included in the chain, or staff—including the firm's experienced outside counsel—failed to follow up. For example, one senior manager flagged the VPIRG answer as a "problematic statement," pointing out that the company could have referred to several underground lines, including the one that was later found to be leaking.[61]

Despite knowing about the "potential inconsistencies" and despite the clear interest in the "underground piping" issue from legislators, activists and the Department, the company never clarified its statements. Especially troubling to Vermont officials was Thayer's commitment in his testimony to do "some research on that and get back to you" and then the lack of subsequent follow-through on this commitment.

Most damning for Entergy, according to the investigators, was the failure of staff to properly brief Site Vice President Colomb before his appearance before the Board. Colomb, the top manager at the plant, was asked the same question as Thayer and gave the same answer, also under oath. As the law firm report stated, "ENVY personnel failed to correct or clarify Thayer's testimony, failed to correct or clarify the ENVY response to VPIRG 4-6 and failed to prepare Mike Colomb so that he could make clear the context of his testimony on May 26, 2009."[62]

Opponents took the opportunity to underscore their narrative of an untrustworthy, out-of-state company. As VPIRG's Duane

Peterson said, "They had to admit they were leaking radiation into the soil from pipes they had earlier said that didn't exist. That's a tough position. We played up our message, the plant was old, dangerous and irresponsible. Old and dangerous were underscored by the thing falling apart. Irresponsible was underscored by their lying. So we just kept hammering on them."[63]

The editorial writers at the *Burlington Free Press*, the state's largest newspaper, provided a window into the emerging dynamic. Prior to 2009, the *Free Press* had not taken a position on relicensing. Published more than 150 miles from Vermont Yankee, the newspaper had not considered the plant an important story until the cooling tower collapse. In December 2009, the editorial writers cited the economic arguments and urged a "rational" debate "on the facts." A month later the paper took a different approach: "The underground pipe incident adds to the noxious drip, drip, drip of Vermont Yankee's troubled history and shows once again that Entergy is its own worst enemy in the plant's bid to win a license extension."

Two weeks later the *Free Press* joined those calling for the plant to close: "Close Vermont Yankee. The time has come for the state to find a new source of energy … Events such as a radioactive leak unresolved more than six weeks after it was first revealed to the public and misinformation provided by Entergy officials under oath raise serious questions about whether Vermont Yankee serves Vermont's long-term interests. These issues make for a strong argument against allowing Vermont Yankee to operate beyond 2012."[64]

With the Senate vote looming, national news media also focused on the debate. Stories appeared in news outlets from coast to coast, speculating about whether Vermont would see the first public vote rejecting a nuclear power plant in 20 years.

Entergy, though, still had some cards to play. The company announced plans to bus in dozens of plant workers to personally deliver the jobs message. And Hébert scheduled a press conference for the day before the vote. Entergy was expected to offer a new power deal. The vote counters in groups opposed to the plant saw the Senate as leaning toward a "no" vote, but not certain, with a few senators still on the fence.[65]

Could a positive power deal change the storyline?

Notes

1 Governor Jim Douglas, "State of the state address," January 7, 2010, http://www.cctv.org/watch-tv/programs/gov-douglas-state-state-address.

2 Tony Klein, Robert Dostis and James Moore, in discussion with the author, May 11, 2011, May 27, 2011, and April 18, 2011, respectively.

3 RKS Research and Consulting, "GMP's Vermont Yankee relicensing survey," August 7, 2009.

4 John Curran, "State, Vermont Yankee reach deal; vote awaits," Associated Press, October 9, 2009.

5 Vermont Public Service Board, "Reply brief of Entergy Nuclear Vermont Yankee, LLC, and Entergy Nuclear Operations, Inc.," Docket 7440, August 7, 2009, http://psb.vermont.gov/sites/psb/files/docket/7440VT_Yankee_Relicensing/EN%20Relicensing%20%20Reply%20Brief%208-07-09%20(Searchable%20PDF)%20FINAL.PDF, pp. 1-4; Bob Audette, "Entergy: Fear should not be PSB' guide," *Brattleboro Reformer,* August 13, 2009.

6 David Gram, "Vermont: Senate says decommissioning must be full before Vermont Yankee nuclear sale," Associated Press, March 21, 2008; Terri Hallenbeck, "Nuclear plant relicensing decision should go ahead, Douglas says," *Burlington Free Press,* January 22, 2010.

7 State of Vermont, "Act 160. An act relating to a certificate of public good for extending the operating license of a nuclear power plant," May 18, 2006, http://www.leg.state.vt.us/jfo/envy/ACT160.pdf.

8 David Lochbaum, in discussion with the author, December 20, 2012.

9 David Gram, "Senate advances bill calling for Vt. Yankee inspection," Associated Press, March 12, 2008.

10 State of Vermont, "Act 189. An act relating to a comprehensive vertical audit and reliability assessment of the Vermont Yankee nuclear facility," Section 3, http://www.leg.state.vt.us/jfo/envy/ACT189.pdf, p. 4.

11 Associated Press, "A quarter of U.S. nuclear plants leaking," February 1, 2010.

12 Those concerns had spurred the New England Coalition to petition the NRC to allow the group to test soil for radioactive contamination back in 2003. In a forerunner to the debates to come, Entergy argued that only the NRC could test the soil, because of their role overseeing safety. The Coalition won permission to conduct the tests but the soil came back clean. Associated Press, "Anti-nuke group can test soil at Vermont Yankee," December 23, 2003.

13 Associated Press, "Douglas administration hits nuke panelists," July 2, 2008.

14 Former PSB Chair Michael Dworkin called this a "stunning example of disrespect for a very expert, widely recognized panel representing the Legislature and the Government." Michael Dworkin, in discussion with the author, January 20, 2012.

15 Nuclear Safety Associates, "Reliability assessment of the Vermont Yankee nuclear facility," prepared for the Vermont Department of Public Service, December 22, 2008, http://publicservice.vermont.gov/dockets/7440/Reliability%20Assessment%20VY%20Final%20-%20REDACTED%20(2).pdf.

16 State of Vermont, "Act 189," p. 4.

[17] Morgan, Lewis & Bockius, LLP, "Report of investigation Entergy Nuclear Vermont Yankee," February 22, 2010, http://pbadupws.nrc.gov/docs/ML1016/ML101650132.pdf (accessed September 29, 2011), p. 34.

[18] David Gram, "Report says Vt. nuke plant can last past 40 years," Associated Press, December 24, 2008. Note: The report did make a series of recommendations to improve reliability, which became "action items" for Entergy to pursue.

[19] Nuclear Safety Associates, "Reliability assessment of the Vermont Yankee nuclear facility," p. 262.

[20] Ken Picard, "The insiders," *Seven Days*, February 17, 2010, http://www.7dvt.com/2010arnie-gundersen-nuclear; Fairewinds Associates, "Who we are," http://fairewinds.com/content/who-we-are.

[21] Tony Klein, in discussion with the author, May 11, 2011. Note: Senate President Peter Shumlin in a speech on the Senate floor on Feb. 24, 2010, had this to say about Gundersen: "He has been an ardent watchdog of the industry who happens also to be the person that knows more about nuclear power plants than anyone else I've met in the state of Vermont. Who also happens to be right, the only person in the state of Vermont who's been right on every single prediction of what might happen at Vermont Yankee. But he took a lot of heat. And I appreciate his service."

[22] Arnie Gundersen, in discussion with the author, August 24, 2011.

[23] Morgan, Lewis & Bockius, LLP, "Report of investigation Entergy Nuclear Vermont Yankee," p. 106; Arnie Gundersen, August 24, 2011.

[24] Morgan, Lewis & Bockius, LLP, "Report of investigation Entergy Nuclear Vermont Yankee," p. 107.

[25] Morgan, Lewis & Bockius, LLP, "Report of investigation Entergy Nuclear Vermont Yankee," February 22, 2010; Richard Smith, in discussion with the author, May 9, 2011.

[26] Morgan, Lewis & Bockius, LLP, "Report of investigation Entergy Nuclear Vermont Yankee," pp. 105-108.

[27] David Gram, "Vt. Yankee says didn't mean to mislead lawmakers," Associate Press, January 14, 2010; Morgan, Lewis & Bockius, LLP, "Report of investigation Entergy Nuclear Vermont Yankee."

[28] Arnie Gundersen, August 24, 2011.

[29] "John Marshall among 34 'groundbreaking lawyers' in the US," and "John H. Marshall profile," Downs Rachlin Martin, PLLC, http://www.drm.com.

[30] Michael Dworkin, in discussion with the author, January 20, 2012.

[31] The VELCO Northwest Reliability Project cost $10 million to litigate through the Public Service Board, see "Planning for Power: Citizen participation in the siting of a high-voltage transmission line in Vermont," (Richard Watts, Doctoral dissertation, University of Vermont), February 2006.

[32] John Dillon, "Following disclosure, Yankee's record is examined," Vermont Public Radio, January 18, 2010, http://www.vpr.net/news_detail/86927/; Vermont Office of the Attorney General, "Vermont Yankee criminal investigation report," July 6, 2011, http://www.atg.state.vt.us/assets/files/Office%20of%20the%20Attorney%20Generals%20Criminal%20Investigation%20Report%20on%20Vermont%20Yankee.pdf.

[33] Anne Galloway, "Leonard pulls Thayer, says shutting down Yankee wouldn't hurt Entergy's bottom line," *VTDigger.org*, February 3, 2010, http://vtdigger.org/2010/02/03/leonard-pulls-thayer-says-shutting-down-yankee-wouldnt-hurt-entergys-bottom-line/.

[34] Morgan, Lewis & Bockius, LLP, "Report of investigation Entergy Nuclear Vermont Yankee, pp. 73-80; Vermont Office of the Attorney General, "Vermont Yankee criminal investigation report," p. 5; Paul Burns, in discussion with the author, June 1, 2011.

[35] Bob Audette, "Tritium leak found at VY," *Brattleboro Reformer*, January 7, 2010; David Gram, "Vt. Yankee well tests show radioactive isotope," Associated Press, January 8, 2010.

[36] Bob Audette, "New well has higher levels of tritium," *Brattleboro Reformer*, February 2, 2010. Note: 2011 water samples taken from the nearest groundwater monitoring well, GZ-10, have shown decreasing tritium concentrations, and water samples taken from wells further east have shown increasing tritium concentrations. Wells GZ-7, GZ-15 and GZ-21 have shown tritium concentrations rising most quickly since the leak was stopped. See map of monitoring well sites here: http://healthvermont.gov/enviro/rad/yankee/documents/VY_tritium_map.pdf and monitoring well concentration graphs here: http://healthvermont.gov/enviro/rad/yankee/documents/VY_Tritium_well_concentration_graphed.pdf.

[37] NRC, "Tritium, radiation protection limits, and drinking water standards," February 2011, http://www.nrc.gov/reading-rm/doc-collections/fact-sheets/tritium-radiation-fs.pdf.

[38] Jeff Donn, "Radioactive tritium has leaked from three-quarters of U.S. nuclear plants," Associated Press, June 21, 2011; U.S. Department of Energy, Argonne National Laboratory, "Human health fact sheet, Tritium (Hydrogen-3)," August 2005, http://www.evs.anl.gov/pub/doc/tritium.pdf.

[39] Allan Mazur, *True Warnings and False Alarms* (Washington, DC: Resources for the Future, 2006).

[40] David Gram, "Vt nuke plant leaks renew debate over aging plants," Associated Press, February 2, 2010.

[41] David Gram, "Hodes: Give NH more oversight over Vermont Yankee," Associated Press, February 9, 2010.

[42] Charlie Donaldson to Jacklyn Brilling, New York Office of the Attorney General, February 10, 2010, http://documents.dps.state.ny.us/public/Common/ViewDoc.aspx?DocRefId=%7B87213914-C199-4337-9523-0506B54E4AD7%7D.

[43] Morgan, Lewis & Bockius, LLP, "Report of investigation Entergy Nuclear Vermont Yankee."

[44] John Dillon, "Yankee admits underground pipes could be source of leak," Vermont Public Radio, January 15, 2010, http://www.vpr.net/news_detail/86897/; David Gram, "Vt. Yankee says didn't mean to mislead," Associated Press, January 14, 2010.

[45] Bob Kinzel, "Governor hopes Yankee will still be part of Vermont's energy future," Vermont Public Radio, January 28, 2010, http://www.vpr.net/news_detail/87042/s.

46 David Gram, "Gov. to lawmakers: Hold off on Vermont Yankee vote," Associated Press, January 28, 2010.

47 Morgan, Lewis & Bockius, LLP, "Report of investigation Entergy Nuclear Vermont Yankee," p. 73.

48 David Gram, "Top Vermont Yankee official 'relieved of duties,'" Associated Press, February 3, 2010; "Entergy corporation outlines steps to restore trust of Vermonters," VTDigger.org, February 4, 2010, http://vtdigger.org/2010/02/04/entergy-corporation-outlines-steps-to-restore-trust-of-vermonters/.

49 Brattleboro Reformer, "Douglas calls for transparency, changes at VY," January 27, 2010.

50 News articles database (1,409 articles). Words searched: accountability, bogus, deceit, deceitful, deceptive, dishonest, fake, false, hid, hide, hiding, honesty, lying, lie, misleading, misled, mistrust, not trustworthy, transparency, lies, trust, honest, liar, distrust, untrustworthy.

51 This analysis was applied to 346 Associated Press news articles written between 2002 and March 1, 2010. For details on the analysis, approach and methods, see Richard Watts; "The role of media actors in reframing the media discourse in the decision to reject relicensing the Vermont Yankee nuclear power plant," Journal of Environmental Studies and Sciences, February 2012.

52 Brattleboro Reformer, "Douglas calls for transparency, changes at VY," January 27, 2010.

53 Curt Hébert, email to J. Wayne Leonard, May 6, 2010.

54 Curt Hébert, "Public profile," http://www.linkedin.com/pub/curt-hebert/31/15b/29a.

55 Curt Hébert press conference, February 22, 2010; Vermont Office of the Attorney General, "Vermont Yankee criminal investigation report;" Morgan, Lewis & Bockius, LLP, "Report of investigation Entergy Nuclear Vermont Yankee."

56 Curt Hébert press conference, February 22, 2010; Entergy, "Independent investigation report commissioned by Entergy turned over to Vermont Attorney General," February 24, 2010, http://www.entergy.com/news_room/newsrelease.aspx?NR_ID=1697.

57 Bob Stannard, in discussion with the author, May 11, 2011.

58 Steve Terry, in discussion with the author, April 18, 2011.

59 The report was publicly released by the Attorney General's Office in April 2010. The attorney general's investigation into criminal behavior was released in July 2011, ultimately not finding enough evidence to prosecute criminal intent.

60 Morgan, Lewis & Bockius, LLP, "Report of investigation Entergy Nuclear Vermont Yankee."

61 Henry Metell email to David Mannai, May 23, 2009. Note: This email is dated at 10:07 p.m. on Saturday, May 23. Vermont Office of the Attorney General, "Vermont Yankee criminal investigation report," July 6, 2011, http://www.atg.state.vt.us/assets/files/Office%20of%20the%20Attorney%20Generals%20Criminal%20Investigation%20Report%20on%20Vermont%20Yankee.pdf.

62 Morgan, Lewis & Bockius, LLP, "Report of investigation Entergy Nuclear Vermont Yankee," p. 13.

63 Duane Peterson, in discussion with the author, April 6, 2011.

[64] Editorial, "Sticking to the facts best for Vt. Yankee debate," *Burlington Free Press,* December 4, 2009; Editorial, "Reasons to question Vt. Yankee's future," *Burlington Free Press,* January 17, 2010; Editorial, "Vermont Yankee no longer an asset," *Burlington Free Press,* Feb 21, 2010.

[65] Paul Burns, in discussion with the author, June 1, 2011.

SIX

Public Fallout

I n the early summer of 2009, Vermont Yankee opponents from
across the state met to plan strategy at a farmhouse outside of
Springfield, some 40 miles north of the plant. Hosted by Denis
and Betsy Rydjeski, members of the local Sierra Club chapter, the
meeting included organizers from VPIRG, the Citizens Awareness
Network and the Toxics Action Network as well as the longtime
citizen activists from the Brattleboro area and a recently arrived
staffer from Greenpeace. The Greenpeace organizer was struck
by how well the group worked together despite the differences
in ages and life experiences. And it was the younger crowd—
the 20-something organizers from Montpelier—who drove
the conversation and doled out tasks, because of their intimate
knowledge of the Vermont State Legislature, the campaign's primary
target.[1]

Organizers wanted a clear vote from the Legislature rejecting the
plant. Technically, under Act 160, no vote at all would have the same
effect. But the organizers and observers knew that a "no" vote would
have tremendous political value. As former PSB Chair Michael
Dworkin said, "Although legally a no vote meant nothing, in the real
world of concerns, impressions, impacts and expectations, a no vote
would be of dramatic importance."[2]

Organizers handed out copies of the coalition's first scorecard,
a list of legislators giving their ranking on a 1 to 5 scale. A "1"
would apply to Rep. David Zuckerman (P-Burlington), an organic

vegetable farmer and activist from Burlington, completely and publicly opposed to Vermont Yankee's continued operation. The "3"s were on the fence—they could fall either way. A "5" would be someone like Rep. Patty O'Donnell (R-Vernon), a leader in the Republican Party and a champion of the plant. Information about legislators' scores was later updated, based on information coming in from community organizing, legislative meetings and other sources. Information was fluid: "I heard so-and-so legislator told this person at the farmers' market that ..." Organizers agreed to only move a legislator's number if they had three independent sources verifying a shift in position.

The opposition groups joined together under the banner of Safe Power Vermont. The Toxics Action Center, an organization that shares an office with VPIRG in Montpelier, hosted weekly conference calls and coordinated an email list where activists could share strategy and compare notes. Jess Edgerly, a charismatic 20-something Blackberry-wielding organizer, took the lead, arranging the weekly strategy calls and keeping the minutes. With a mop of red hair and a bright smile, Edgerly was a strong bridge between the longtime citizen activists and the younger Montpelier-based professionally trained organizers.

Since the passage of Act 160 in 2006, the plant's opponents had focused on the 150 members of Vermont's House of Representatives. Activists thought they would have a greater impact there, because House members each represent about 4,000 Vermonters. The state's 30 senators represent districts ranging from 15,000 to more than 100,000 in Chittenden County. Five or six letters or calls for a House member might have a real impact. For a senator, five or six calls might not have much influence.[3]

Through the summer, VPIRG and Citizens Awareness organizers went door to door, talking directly with Vermonters on porches and in living rooms and kitchens across the state. For VPIRG it was the third year canvassing on Vermont Yankee, gathering signatures and donations, a focus that had helped the organization to achieve record summer fundraising totals. The 2009 summer canvass added a new twist, with door-to-door campaigners generating picture postcards. Photographed at the front door or in the yard, families and individuals stood in front of a "Repower VT: Replace VY" sign and asked their legislators to vote to close Vermont Yankee. That fall, VPIRG mailed every legislator a disk with the photos of the families from their district—more than 6,000 photos in total.[4] As one legislator said, "That was very powerful. Because people looked at these pictures and said, wow, he thinks that. He's that good old guy from down at the general store."[5]

Figure 19. "Repower VT: Replace VY" summer canvass photo postcard, 2009. Courtesy of VPIRG.

Canvassers are a persistent group. One canvasser, for example, had parents who were friendly with the Douglas family (that's Governor Douglas). When her daily route took her near the Douglas home in Middlebury, she'd stop and knock on the door. One day the governor was home and she delivered her "rap"—the core message each canvasser delivers: "Our state could lead the country, showing that green power will create good jobs and protect our environment … But, it won't be easy. Vermont Yankee wants to lock the state into two more decades of risk and radiation …" Douglas listened from behind his screen door but politely refused to sign the petition.[6]

At the end of the summer, VPIRG kept the canvassers on the payroll for some direct Vermont Yankee advocacy. After a summer of talking with Vermonters in rain and sun, the mostly college-aged canvassers are a formidable force. Often from Vermont, they are bright, motivated and knowledgeable about the state and state policies as well as expert at all the social interactions that enable them to raise more than $160 a night. VPIRG sent teams of these canvassers directly into towns to generate local letters and calls to those legislators identified as being on the fence on the Vermont Yankee issue.

One success story was with Rep. Frank Geier (D-South Burlington). South Burlington is a sprawling suburban town with a core of old neighborhoods and multiple new developments of large single-family homes and condominiums. One of Vermont's fastest-growing towns (pop. 16,000), South Burlington is the fourth-largest city in the state. Traditionally less liberal than Burlington, the city has excellent schools and a thriving retail center. It's a very long way from Vermont Yankee, geographically, socially and culturally. Debates at city council and school board meetings and around the ice at Cairns Arena focus on issues of the city's growth and budgets, services and school quality. Geier was on the fence, a "3" in the campaign's targeting numbers. VPIRG sent teams of canvassers into Geier's neighborhood, requesting letters, emails and calls. As the canvassing director recalls it, "After a few days canvassers had something like 15 personal letters asking him to do the responsible thing. I remember talking with the president of the condo complex and she knew him personally and wrote a letter. My eighth-grade English teacher lived in the condo complex and she wrote him a letter. And he went from a 3, to a week or two later, walking into the VPIRG office, he spoke with the first person he saw and said 'I'm with you on VY, what can I do to help?' It was night and day."[7]

In another story, the organizers arranged a special event to focus on Rep. Adam Greshin (I-Waitsfield), an independent representing the artistic, recreational and politically active Mad River Valley community. The valley is home to two of Vermont's most celebrated ski areas, the 4,000-acre, 111-trail Sugarbush resort and its alter ego, Mad River Glen—the only cooperatively owned ski area in the country and one of the last two ski areas in the U.S. still running a single chair. Greshin was a co-owner of Sugarbush and a prominent businessman who had voted against the decommissioning bill. The organizers arranged for a "Spank-You Adam Greshin" event at American Flatbread—a local gathering place. All afternoon and evening, people came by and signed the "Spank-You" card. Shortly afterwards, Greshin changed from being on the fence to opposing the plant. As an organizer explained it, "I think for him more than anything else it was just political. Being neutral on the plant was being politically stupid. And so he decided to change."[8]

By the fall of 2009, coalition vote counters believed they had close to enough votes, somewhere around the 76 needed for a majority.[9] But there was no indication from House Speaker Shap Smith that he would bring the issue to the members. Furthermore, even if the speaker did schedule a vote, the organizers were not confident in the outcome. Despite strong support in the House, at the end of the day a hotly contested vote like this would require the speaker and his leadership team to personally engage. Smith gave no indications that he was willing to conduct the arm-twisting needed to guarantee success. In the previous session, Smith and House leaders had worked extremely hard to win an override of Douglas's veto of the 2008 gay marriage law. It would take that kind of effort again to win the proposed "no" vote on Vermont Yankee.

One fall day, VPIRG's Paul Burns and James Moore drove down to Putney to meet with Senate President Peter Shumlin at his home in the shadow of Putney Mountain, past apple orchards brimming with fruit. Shumlin asked them to switch the campaign's focus from the House to the Senate. Organizers and legislative allies were at first skeptical. Said Shumlin's fellow Windham County Senator Jeanette White, "Peter and I had some conversations about it. I really questioned him at first. Because I didn't think we had it, because we had some pretty conservative people in the Senate at the time. This was a big vote. And you couldn't get this one wrong. You had to win and you had to win the right way."[10] Although only 16 votes were needed, senators tended to be harder to influence and more "business oriented" than House members.[11]

As Burns said, "Knowing some members of the Senate, we felt this would be tough. These were not anti-nuke people by any stretch and not particularly prone to going with an environmental argument over a business argument. It's not an easy thing to really shift your focus. But ultimately when you have a figure as persuasive and powerful and well-positioned as Peter Shumlin was, you have got to listen." At the time, Burns at best could count only 12 or 13 likely "no" votes and thought it would be a real struggle to get to 16. Still, Shumlin was convincing. And his personal involvement would make a difference. VPIRG in turn convinced the other groups in the coalition to shift the campaign's focus from the House to the Senate.[12]

Through the late fall of 2009, the campaign continued to build momentum. Activists working closely with a small group of legislators generated phone calls, emails and letters and met individually with legislators to press the case for plant closure. With the Senate as the target, activists focused more and more on individual senators. Rather than blanket calls or email blasts, organizers coached individual callers to hit the right message, a message that would influence that individual senator. As one organizer explained it, "We decided to really up the customization. We'd steer it a little bit from a messaging standpoint because we didn't want people flying off the handle. But mostly we were writing down their thoughts and giving it back to them. And we were also getting these folks to call us back and say how the phone call went so we could refine our targeting lists and have a better sense of who we needed to focus on. It was a couple years worth of organizing in a couple months."[13]

The organizers were also increasingly trying to create an environment in which voting for Vermont Yankee would be difficult. To do this they had to bring the campaign to places where Vermont Yankee was not top of mind. At the end of day, it would be legislators in districts hundreds of miles from the plant who would determine the outcome. Speaking metaphorically, it is hard to get much further from Vernon than the Lake Champlain shore town of Shelburne, Vermont's wealthiest town. Annual household income in Shelburne averages $125,000, more than twice that of

Vernon ($51,000).[14] Spread along Lake Champlain, five miles from Burlington, Shelburne is home to many of Vermont's political and business elite, including utility executives, former governors and at least one billionaire.[15]

In September 2009, Entergy's sponsorship of the Shelburne Farms Harvest Festival, a celebrated local event that draws thousands of visitors, attracted activists' attention. Established as an experimental farm by the daughter of Cornelius Vanderbilt in the 1800s, Shelburne Farms is a breathtakingly beautiful, landscaped and turreted, half farm and half medieval fortress on a bluff overlooking Lake Champlain. A dozen volunteers, all dressed in suits and Mr. Burns masks (a reference to the villainous nuclear plant owner on the television series, *The Simpsons*), descended on the bemused families and visitors on a splendid fall day. As visitors watched wool being spun from the farm's specially bred sheep, organizers decorated the portable toilets with signs that read, "This port-a-potty leaks less than Vermont Yankee. Why? Call your State Rep. to find out!"[16] Organizers were everywhere, in parades, at events across the state, using humor and style to tell the story of an aging power plant that should be retired (see *Figure 20*).

Working closely with Shumlin and other legislators, activists checked in regularly on vote totals, adjusting the legislators' scores on the 1 to 5 scale as information filtered in. By the late fall, organizers saw the election in the Senate as close, but winnable. Still it was well short of the large margin needed to score a lasting victory. More needed to be done. To move the campaign to the next level, the one thing really missing was money.

Momentum Builds

Ten days after Entergy Vice President Jay Thayer filed his "best offer" letter with the Public Service Board, VPIRG's James Moore met with a small group of potential financial supporters in the wood-paneled offices of Seventh Generation on Burlington's waterfront. Moore was joined by VPIRG Board Chair Duane Peterson, a self-described organizing entrepreneur with experience in more than 17 political campaigns. As Peterson tells the story, "We started

Figure 20. A float designed to look like an aging nuclear power plant is pulled by a 1972 Pontiac LeMans—to underscore the age of Vermont Yankee—during the 15th Annual Magic Hat Mardi Gras Parade in Burlington. Photo by Glenn Russell, *Burlington Free Press.*

kicking it around. And, in a stream of consciousness, I laid out what campaigns look like and how this was a golden opportunity. We're going up against an $11 billion company. But we actually have a shot. Never in its 40-year history has Vermont Yankee been so vulnerable. Let's do this."[17]

Such a campaign would take market research, it would take polling and it would take paid media. All of this would cost money. Following the meeting, Moore walked one of the potential donors to the donor's car. It was a bitter cold December night with light snowflakes falling. The donor pulled out a wallet and wrote a check for $100,000. "Can you get to a bank tomorrow?" the donor asked. It was the last day of the calendar year.

With the surge in resources, Peterson joined the campaign full-time, sharing responsibilities with VPIRG's Moore. Moore focused on the legislative strategy while Edgerly and VPIRG's Ben Walsh, a calm, patient and disciplined organizer, coordinated the grass-roots efforts. Often in blue jeans, open-collared oxfords and a blue blazer, the 50-something Peterson added professional political experience to the campaign. Peterson had moved to Vermont about 10 years before to manage the progressive enterprises of Ben Cohen and Jerry Greenfield (a.k.a. Ben & Jerry, of ice cream fame), working with them to found TrueMajority.com. In January, Peterson tacked a 1975 news article headlined "VPIRG Wants Plan for Closing Yankee" above his desk in the cramped staff advocate's office and went to work.[18]

The campaign funded a public survey to test messaging strategies and take a snapshot of Vermont opinions. Even by mid-January (the survey was conducted January 19-22) Vermonters remained split on Vermont Yankee, with about one-third unfavorable, one-third favorable and one-third neutral/no opinion. And on the key question of relicensing Vermont Yankee, Vermonters split evenly: 44 percent against relicensing and 42 percent in favor. However, Vermonters were sharply negative on the spin-off plan, with more than half opposed and less than 10 percent in favor.

These results showed a growing shift toward public support for closing the plant in 2012 when compared to results from a poll conducted for Green Mountain Power only six months before (July 21-29, 2009). Before the tritium leaks and the headlines about misleading testimony, 42 percent of GMP residential customers supported relicensing Vermont Yankee, 24 percent were opposed and 29 percent were neutral or not sure. The Green Mountain Power pollsters noted that 80 percent of the neutrals could be switched to supporting relicensing if supporters focused on the plant's economic benefits.

By mid-January, many of those neutrals had moved to opposing the plant, but overall public opinion remained evenly split. The organizers parsed through the results, looking for ideas for tactics and messages. For example, there was a large gender divide, with women (48 percent) much more likely than men (38 percent) to oppose the license extension. And there was a significant gap between Democratic voters, who opposed the plant by a margin of 33 points and Republicans, who supported the plant's relicensing by a margin of 49 points.[19]

The poll confirmed organizers' suspicions that economic and jobs messages were big winners for Entergy. Avoid talking about jobs, the pollsters advised, and instead always pivot back to the core message: "Closing Vermont Yankee *as scheduled* is the safe and responsible thing to do." Pollsters summarized the most persuasive arguments as:

1. The plant is too old to be safe or reliable,
2. The owner's out-of-state corporate executives are untrustworthy, and
3. Relicensing the plant means huge profits for the corporation, but Vermonters could be stuck with the cleanup bill.

The pollsters crystallized the most potent message, "Vermont Yankee is simply not safe, and we can't trust them to tell us the truth about it." The pollsters also tested message sponsors, checking various categories of Vermonters and Vermont organizations. Not surprisingly, by January 2010, "out-of-state utility executives" were the least-popular category of people in the state. "Environmental activists" fell about in the middle, right next to Vermont Yankee employees. Farmers were considered the most popular and trustworthy.[20]

Peterson and a television crew immediately spent a day with a local farmer, filming him walking past old farm machinery, mucking out his barn and talking conversationally to the camera: "So on the farm I've got a lot of pieces of equipment. And as the equipment gets old and unreliable and unsafe, I retire it. It's the same thing with a nuclear reactor. As it gets old and unreliable and unsafe and it's leaking radiation, it's time to retire it. And I do not want to get stuck with the cleanup bill. It's common sense to retire it on schedule. And it's the right thing to do." [21]

Senate Vote Looming

The opposition organizers now had the structure of an effective campaign in place. A campaign plan for Safe Power Vermont detailed an aggressive list of goals for letters to the editor, events, emails and phone calls to be completed by the end of the legislative session in late April or early May, when the vote was expected. When the news broke about Entergy's misleading testimony, that timeline was accelerated. Moore immediately emailed campaign supporters across the state: "Reports are that Entergy lied under oath. We have to make sure legislators know that Vermonters won't stand for this kind of corporate behavior any longer."[22]

With several key senators still undecided, the news about Entergy's misleading statements had an immediate impact. Ann Cummings, chair of the Senate Finance Committee, wrote to a constituent on January 15: "The recent revelations about the underground pipes at VY have confirmed my growing suspicion that Entergy is not the kind of corporation that Vermont wants to do business with. I'll do everything I can to make sure that VY closes on time."[23] Organizers moved Cummings to a "1" on their scorecard.

Dairy farmer and Addison County Senator Harold Giard also moved to a certain "no," writing "I will vote to shut VY down the first chance I get ... Vermonters should not have to sleep with 'one eye open' for the next twenty years as that facility limps along, held together with bailing twine, Elmers glue, and 'hope.'" [24]

On January 19, Greenpeace sent a second full-time organizer to the state to focus on Rutland County and central Vermont. Meanwhile, letters to the editor had doubled from the previous week. The campaign was hitting full stride. Thousands of calls and emails were being generated to various legislators.[25] A group marching from Brattleboro, through snow and ice, arrived at the State House after a 12-day, 126-mile journey. Marchers slept in church basements and at strangers' houses. The two campaigns were coming together. Citizens who had been opposing the plant for more than 30 years joined the Montpelier-focused activists at the State House. Along the way, the southern Vermont-area marchers were joined by supporters from across the state. By the final day, the marchers' numbers had swelled to 200.[26]

The campaign team met frequently, brainstorming creative tactics to keep bringing home the story of an aging plant run by untrustworthy owners. Their job was to take a "megaphone" to Entergy's missteps and promote their chosen storyline. Each day brought a new opportunity.[27] Peterson, for example, bought Entergy stock, allowing him to join Entergy's financial status conference calls. Organizers tracked down and briefed several of the analysts prior to the company's January call. Sure enough, an analyst asked about the "resistance" in Vermont and the effect of Vermont Yankee's possible rejection on Entergy's stock. As Peterson tells it, he almost fell out of his chair when Entergy's CEO, J. Wayne Leonard, said, "closing the plant would be cash neutral for the parent corporation because it's not making money for us now… it really wouldn't matter if we had to shut it down." To Peterson, Entergy's story in Vermont had been that the plant was critical for Vermont, an economic engine and a job creator. "And here the guy was saying no big deal and we'll just throw it under the bus."[28] A transcript was released to the Vermont press, generating another day of negative stories about Entergy.

Crisis Grows for Entergy

Meanwhile, at Vermont Yankee, the identification of underground pipes as the source of the leaks accelerated the company's troubles. Entergy officials scrambled to contain the damage, both at the plant and in the growing firestorm of public criticism. At the plant, staff in the chemistry department charged with finding the source of the leak basically worked around the clock for 40 days, opening a "war room" and drawing experts from across the company. As they worked, the negative press swirled around them. By the end of January, at least five investigations were underway, along with a lawsuit calling for the plant to immediately close. For Vermont Yankee employees, this was an extremely difficult time.

As momentum built for a vote in the Legislature, Governor Douglas called for a "time out," saying that although he remained a supporter of the plant and thought trust could be regained, it was time to pause before the Legislature rushed to a decision. Let the various investigations finish before voting, Douglas said. Legislative leaders disagreed. Activists called the statement an act of "desperation."[29]

Plant opponents sponsored a press conference to underscore the storyline of Entergy as an untrustworthy company and to attack the company's "I AM VY" ads. At the press conference VPIRG released a list of Entergy's "lies and deceptions" and demanded further criminal investigations, pointing out that some of the senior managers under investigation were in the ads. For example, one of the employees featured in ads about trust and value, was now under investigation for misleading statements. A few days later, Entergy pulled all of its ads off the air. The Associated Press's David Gram later used the talking points from the press event as a hook for a news story on Entergy's past "deceptions," some of which dated back to when Entergy purchased the plant. At the same time, the Southern Vermont group Safe & Green sponsored a "Leaks and Lies" rally and parade in downtown Brattleboro.[30]

Entergy's Vice President Curt Hébert, now in Vermont almost full-time, worked with Entergy's lobbyists and Douglas to slow down the opposition's momentum, suggesting that additional information was needed before a vote. Legislative allies asked the Senate to delay voting until the results of the investigations were in. Business leaders urged the Senate to go slow. "This is not a responsible direction the Senate is taking," said IBM's John O'Kane.[31]

Throughout January and February the story continued to grow, expanding from a state story into a national story. National media focused in on Vermont's "unique" role in deciding the plant's future. By February the story had appeared in hundreds of media outlets outside of the state, including the *New York Times*, *Wall Street Journal*, ABC News, CNN and the *Boston Globe*.[32] The *New York Times'* Matthew Wald suggested that "a particularly adamant cadre of environmentalists who oppose nuclear reactors has gained substantial momentum in the last few weeks …"[33]

The increased press attention was accompanied by added scrutiny from Congress and federal officials. In Washington, congressmen from New York, Massachusetts and New Jersey introduced a bill to investigate "leaky pipes at nuclear plants."[34] President Obama's appointees to the Nuclear Regulatory Commission were asked in a Congressional hearing about the leaking tritium at Vermont Yankee. "The point is not that it's hurting anyone. The point is that it's showing that you don't have your act together," nominee William Magwood, IV, told the U.S. Senate's Environment and Public Works Committee.[35]

Vermont organizers requested that the U.S. Justice Department investigate Entergy's actions. Congressmen from neighboring states asked that their states also have an oversight role in the plant's future.[36] And regulators in New York were given a copy of the law firm's report on the internal investigation, a report that showcased the multiple internal errors by Entergy staff. These New York regulators had oversight over Entergy's four merchant reactors in New York and were still considering Entergy's proposed spin-off of those reactors into a new holding company. For Entergy, putting the report in their hands raised the company's "risk" level.[37]

Retire as Scheduled

Centered at VPIRG's Montpelier office, a few blocks from the State House, the campaign to close Vermont Yankee continued to build momentum. Staff met every morning to review daily plans and adjust strategy, also meeting frequently with Shumlin and other legislators to check vote counts. By now more than five VPIRG staff were engaged full-time in the effort to close the plant. Organizers from Greenpeace, League of Conservation Voters, Vermont Natural Resources Council, the Toxics Action Center, Citizens Awareness Network and citizen activists with Safe Power Vermont joined the strategy sessions and mapped the final weeks of the campaign.

One of the campaign's biggest tasks was keeping plant opponents on message. Take for example the concept of framing the issue as closing Vermont Yankee "as scheduled" instead of calling for a plant "shutdown." Through this emphasis, organizers highlighted the 2012 closure as the status quo, framing the 20-year license extension as unusual, unexpected and unplanned. As Peterson explained it: "One of the messaging points was that Entergy is trying to change the rules of the game. It's a special interest that's coming in here and trying to renew a license that is set to retire for some really good reasons. We needed to make it reasonable, acceptable and mainstream, that this thing was going to retire on schedule."[38]

Or as Senator White, the Windham County legislator, said, "It puts in people's minds—again—that picture, as scheduled, it's always been scheduled. It hasn't been a surprise. They've had 40 years to plan. It's not as if suddenly we threw a loop, because they'd known about it for 40 years."[39] Activists shared poll results with legislators and supporters to encourage consistent messaging, to keep to the storyline of an aging plant at the end of its planned life.

The effectiveness of this strategy can be seen in the growing appearance of this phraseology in press stories. Here for example is Shumlin speaking to a reporter in February: "We need to plan for Vermont's energy future and our businesses, our taxpayers, our ratepayers and the workers of Vermont Yankee deserve to know whether we're going to *retire the plant on schedule* in 2012" (italics added).[40] Reporters also picked up this word formulation, with WCAX reporter Kristin Carlson describing the legislation as about "extending the license for the nuclear plant beyond its scheduled March 2012 closure."[41] By 2010, words associated with the "retire as scheduled" concept had increased five-fold over 2004, from 50 appearances in 2004 to more than 250 in 2012. Activists created campaign materials promoting this phrasing.

Figure 21. Two activists hoist a "Retire Vermont Yankee as planned" banner in front of the State House in February 2010. Photo by Roger Crowley, courtesy of VPIRG.

In mid-February, Shumlin announced plans to bring the question to a vote. The organizers counted 15 certain votes in hand.[42] Moore broadcast the news to activists across the state. "After years of work by thousands of Vermonters, this is it." Staying on message, Moore reminded readers that the plant was too old to be safe and that Entergy was more interested in "profit" than maintenance. The email ended with a call to action: "They are going to throw everything they have at us. We need to do the same. Now is the time to fight."[43]

That evening Shumlin joined a group of environmental activists at a Montpelier bar for the weekly "green drinks" social event. Todd Bailey from the League of Conservation Voters showed Shumlin the results from a just-released WCAX-TV poll. Shumlin leaned in to study the small screen on the PDA. Public opinion had continued to turn against the plant, with 49 percent saying Vermont Yankee should not get the license extension, 27 percent saying they should— and 24 percent not sure.[44]

With a week to go, the activists turned up the heat on the remaining undecided legislators. *Times Argus* reporter Louis Porter surveyed all 30 senators after Shumlin's announcement. Sen. Alice Nitka (D-Ludlow) remained undecided, telling Porter, "I am going to wait and see what the question is. I would like a little more information, frankly."[45] A group of Nitka's constituents scheduled a breakfast meeting with her for that weekend and the organizers went to work. On Saturday morning, cars jammed the parking lot at the Crows Corner Bakery in the tiny Vermont town of Proctorsville. By 9:30, more than 25 people had filled every chair in the café. Nitka walked in the door, looked at the room, and immediately went from being on the fence to a solid "no," leading off her remarks by saying that she would be voting to close the plant.

That same week, Sen. Bill Doyle (R-Montpelier), the dean of the Vermont Senate, was also featured as on the fence. With 40 years of consecutive service, Doyle's legendary skills in walking a tight line between his more conservative inclinations and his increasingly Democratic district were on sharp display. Emails and letters and phone calls poured in. Late on the Sunday evening before Wednesday's vote, Doyle wrote back to a constituent's query: "WILL VOTE TO CLOSE YANKEE" (emphasis in original).

Within 30 minutes Doyle's note was broadcast to the Safe Power Vermont email list.[46]

Entergy's Last Offer

On Tuesday, the day before the vote, Entergy's Hébert held a press conference in the State House's Cedar Creek Room. That morning Entergy had bused in dozens of plant employees. Joined with plant supporters and lobbyists, the employees had roamed the State House for most of the day, buttonholing legislators and debating plant opponents. They then gathered for the press conference, with Hébert at the podium, dressed in a dark suit, blue shirt and purple tie. Hébert, a Louisiana native, spoke of sharing Vermont's values, of a commitment to do business in Vermont, of "good will" towards the state. He discussed economic development, jobs—"650 reasons to support the plant"—and the clean, low-carbon, reliable, low-cost energy the plant generates.[47]

Talking about the numerous discussions he'd had over the last few months, the concerns expressed, the mistakes the company had made and the plans to move forward, Hébert said: "We didn't earn nice. But we're going to earn it. They wanted something different. They wanted—I heard from numerous people—a game changer. Something that makes Entergy and Vermont Yankee look different."

Hébert then announced the "game changer"—a reduced price of $.04/kWh for the first 25 megawatts of electricity sold to Vermont utilities. This was an "unbelievable low rate for electricity," he said. "We need to understand it's a tough economy, and everyone needs to do their part. And so we're going to do our part." Drama was high with a vote scheduled for the next day and the plant's future on the line. Reporters filled the first row of the packed room, some with television cameras. Legislators and activists looked on.

The questions were skeptical.

One reporter asked, "It seems like there's a big condition that has to occur. Utilities have to agree to a PPA [purchase price agreement] that's like what you offered December 18, which they haven't agreed to. Is that not a big 'if' in there?"[48]

Another reporter asked whether the public should be surprised by the leaking tritium. A combative Hébert responded, "It's a manufacturing plant! They don't make garments. They don't make lingerie. They make electricity. And things happen at manufacturing plants. Things break. Things have to be replaced. Things leak. But the public is not in danger. The employees are not in danger ... And I don't know why anyone would have been surprised by that."

Reporters asked about Entergy's communications with the public. In a testy exchange, Hébert pointed out that all information about all incidents at the plant was available to the NRC's resident engineer and DPS staff: "If it's available 24 hours a day to the state engineer and to the NRC 365 days a year, how much more transparent would you like for us to be?" The reporter tried to interrupt, but Hébert continued, "I think *this* is transparent. I think what we're doing here is—you know, I'm answering your questions, we're talking about it. I'm saying we've made mistakes! Look, we're in America, we're in Vermont. Thank *God* people are allowed to make mistakes and move on!"

Hébert then turned to the reporter, saying, "Why would I want to send out a press release on a CR (condition report) every time one is filed?" The reporter answered: "In case you want to keep us informed of what's going on."

As the tone of the press conference degenerated, a reporter for WCAX asked Hébert if this was "some sort of bribe masked as benevolence?" Hébert replied, "Bribes are illegal. I certainly wouldn't enter into that." The reporter said, "Your critics are already saying this is too little too late. What do you say?" Hébert responded, "Too little too late? I don't follow you. You're going to have to help me with that one."[49] Noticeably absent from the press conference were the state's utilities that would actually buy and distribute the low-priced power. Without their participation, there could be no deal.

Moments after Hébert was done, Shumlin stood at the same podium with Senate Majority Leader John Campbell and said the offer did "nothing to change the underlying concerns that we have and our conclusion that Vermont Yankee should be shut down as planned in 2012."[50]

Figure 22. Senate President Pro Tem. Peter Shumlin, D-Windham (center) and Senator John Campbell, D-Windsor, announce plans for a legislative vote on the future of Vermont Yankee. Photo by Glenn Russell, *Burlington Free Press.*

Employees from the plant looked on anxiously, with their jobs on the line. Hébert had said the timing of the press conference had nothing to do with Senate's vote: "Doesn't have anything to do with the legislature, it doesn't have anything to do with anything other than the fact that we want to keep jobs here, we want to help." But the headlines the next day tied the two events together: "On the Eve of Senate Vote, Entergy Offers Discounted Power."[51]

The Vote: Cold and Snow

The night before the scheduled vote it snowed eight inches. Driving was treacherous. Concerned that not enough activists would get in the next day, organizers arranged for people to come early and sleep on living room floors and couches. Organizers were also concerned about conflicts between plant workers and activists arriving from around the state. Messages flew around on the coordinating email

Figure 23. Vermont residents and activists join a rally outside the State House in opposition to Vermont Yankee on February 24, 2010. Photo by Basil Tsimoyianis, courtesy of Greenpeace.

list. If the climate turned hostile and reporters focused on an "environmentalists vs. workers" story, it would overshadow the vote. Attendees were asked to be respectful and not engage directly with plant workers. Citizens Awareness's Stannard urged civility: "We cannot afford to have any outbursts or misbehaving tomorrow. One small incident could piss off one of our tepid supporters and down we go."[52] Another activist wrote, "I will be knitting. I will bring extra needles and yarn for others to participate. That is my calming contribution." Wednesday found the State House packed only with supporters. The Vermont Yankee employees had gone back to work.

Opposition calls continued to pour into the State House all day. Organizers counted 600 calls to the Legislature on the day of the vote, 50 to 150 phone calls for each senator still considered to be on the fence. Sen. Matt Choate (R-St. Johnsbury) from the state's northeast corner—more than a two-hour drive from the plant—was shown on WCAX flipping through the pink slips recording calls from the State House switchboard. "Are you kidding me? How

Figure 24. Activists making phone calls the day of the Senate vote, February 24, 2010. Photo by Roger Crowley, courtesy of VPIRG.

could I vote for anything else?" he said.[53] Because only a small number of the hundreds in the State House could actually squeeze into the Senate's ornate chambers, activists photographed citizen attendees, printed the photos, added hand-scribbled notes and handed them to pages to carry to individual legislators.

Organizers thought they had the votes to win, but the margin of victory was uncertain. And they felt they had to have a strong margin to counter attempts to revisit or discredit the vote. During the two-hour debate, senators ripped the plant and the plant's owners, as citizens, many wearing yellow stickers reading "Retire Vermont Yankee as planned" looked on. After about an hour, Senator Randy Brock (R-Franklin) stood up to speak. A conservative Republican, solid nuclear power supporter and former manager at Goldman Sachs, Brock had only started to become disenchanted in the last few weeks. With obvious unease, Brock announced that he would be voting "no": "The dissembling, the prevarication, the lack of candor—have been striking, and there's not enough time to be able to correct that through management changes or through the kinds of things that we had hoped with time, we could resolve ..."[54]

As Shumlin's ally, Senator White, tells the story, "We had people doing vote counts, every 10 minutes doing another vote count to make sure. When we started the debate—I had like 18 or 19 that I was pretty sure of and I was confident that we would win it. But I had no idea where a lot of people were. And when Randy Brock voted "no" I don't think that the counters knew. I don't think Peter [Shumlin] knew. I think he was stunned by the vote. I think everyone was stunned by the vote. When Randy Brock voted "no," I said, we have it, hands down. It was so clear."[55]

Shumlin concluded the debate by emphasizing issues of trust and competence: "If you can trust them, if they were in fact telling the truth that they didn't know there were underground pipes under the plant, then the obvious question is, well, what's worse? A company that won't tell the truth or a company that's operating an aging power plant on the banks of the Connecticut River and doesn't know that they have pipes with radioactive water running through them, that are leaking, and they don't know because they didn't know the pipes existed. Neither is very comforting."[56] In his speech Shumlin mentioned "retire as scheduled" six times.

Senators voted to close the plant 26 to 4. Onlookers erupted in cheers, hugging and crying.

Notes

1. Jarred Cobb, in discussion with the author, October 3, 2011.
2. Michael Dworkin, in discussion with the author, January 20, 2012.
3. Paul Burns, in discussion with the author, June 1, 2011.
4. Ben Walsh, in discussion with the author, April 18, 2011; Matt Ryan, "Petition opposes Vermont Yankee extension," *Burlington Free Press*, August 14, 2009.
5. Jeanette White, in discussion with the author, June 11, 2011.
6. Jess Edgerly, in discussion with the author, April 18, 2011.
7. Ibid.
8. Ben Walsh, April 18, 2011. Note: For a detailed look at the substance of the campaign that summer see: James Moore, "Repowering Vermont: Replacing Vermont Yankee for a clean energy future," Vermont Public Interest and Education Fund, Summer 2009, http://www.vpirg.org/download/2009-VPIREF-Repowering-VT.pdf.
9. Jarred Cobb, October 3, 2011.
10. Jeanette White, June 11, 2011.
11. Paul Burns, in discussion with the author, June 1, 2011.
12. Paul Burns, June 1, 2011; James Moore, in discussion with the author, April 18, 2011.
13. Jess Edgerly, April 18, 2011.
14. Vermont Department of Taxes, http://www.state.vt.us/tax/statisticsincome.shtml.
15. Francis Storrs, "The 50 Wealthiest Bostonians," *Boston Magazine*, March 2006, http://www.bostonmagazine.com/articles/the_50_wealthiest_bostonians/. Note: John Abele, the owner of the Bostwick Farm property in Shelburne, had wealth in 2006 valued at $3.3 billion. The property is adjacent to Shelburne Farms.
16. Jarred Cobb, October 3, 2011.
17. Duane Peterson, in discussion with the author, April 6, 2011.
18. "VPIRG wants plan for closing Yankee," *Brattleboro Reformer*, March 5, 1975.
19. Grove Insight, "Building support for closing Vermont Yankee: Report of findings from a survey of 600 Vermont voters statewide with oversamples in Addison and Caledonia counties," January 19-22, 2010. Question wording: *Do you think the state legislature should approve or deny Entergy's application to extend the operating license for Vermont Yankee nuclear power plant?*
20. Grove Insight, "Building support for closing Vermont Yankee: Report of findings from a survey of 600 Vermont voters statewide with oversamples in Addison and Caledonia counties."
21. Online at http://www.youtube.com/watch?v=qE8Cls6Ngu4.
22. James Moore, email, January 14, 2010.
23. Ann Cummings, email to Andrea Stander, January 15, 2010.
24. Harold Giard, email, February 8, 2010.
25. Duane Peterson, campaign update, February 5, 2010, and February 15, 2010.
26. Chris Garofolo, "Vermont Yankee supporters, foes descend on Statehouse," *Brattleboro Reformer*, January 12, 2010; Safe and Green Campaign, "Step it up walk," http://www.safeandgreencampaign.org/content/step-it-walk.
27. Bob Stannard, in discussion with the author, May 11, 2011; Duane Peterson, in discussion with the author, April 6, 2011.
28. Duane Peterson, April 6, 2011.

29 Duane Peterson, email, January 27, 2010.

30 Vermont Public Interest Research Group, "Entergy's incompetence and deceptions," January 20, 2010, http://www.vpirg.org/download/Entergy%20 Incompetence%20and%20Deceptions%201%2020%2010.pdf; Philip Baruth, "So busted: VY disappears Dave McElwee," *Vermont Daily Briefing*, January 22, 2010, http://vermontdailybriefing.com/?p=1406; David Gram, "2nd Vt. Yankee well tests positive for isotope," Associated Press, January 20, 2010.

31 Jack Thurston, "Are lawmakers moving too fast on Vt. Yankee?" WCAX, Feb 22, 2010, http://www.wcax.com/story/12026231/are-lawmakers-moving-too-fast-on-vt-yankee.

32 Media database.

33 Matthew L. Wald, "Leaks Trouble Nominees for Nuclear Panel," *New York Times*, February 10, 2010.

34 "Reps. Markey, Hall, and Adler request investigation into leaky pipes at nuclear plants," January 14, 2010, http://markey.house.gov/index.php?option=com_cont ent&task=view&id=3829&Itemid=141.

35 Wald, "Leaks Trouble Nominees for Nuclear Panel."

36 David Gram, "Hodes: Give NH more oversight over Vermont Yankee," Associated Press, February 9, 2010.

37 Curt Hébert, email to J. Wayne Leonard, May 6, 2010.

38 Duane Peterson, April 6, 2011.

39 Jeanette White, June 11, 2011.

40 Bob Kinzel, "Shumlin defends decision on Yankee vote," Vermont Public Radio, February 19, 2010.

41 Kristin Carlson, WCAX, February 22, 2010.

42 Duane Peterson, campaign update, February 5, 2010, and February 15, 2010.

43 James Moore, email, February 16, 2010.

44 Kristin Carlson, "Vermont Yankee: Should it stay open?" WCAX, February 18, 2010, http://www.wcax.com/story/12008043/vermont-yankee-should-it-stay-open; Author notes from attending "Green Drinks" at Black Door Bar and Bistro in Montpelier, VT, February 18, 2010.

45 Louis Porter, "Few Senators back re-licensing," *Times Argus*, February 17, 2010, http://timesargus.com/article/20100217/NEWS01/2170344/1002/ NEWS01.

46 Bill Doyle, email, February 21, 2010.

47 Curt Hébert, press conference, February 23, 2010.

48 Ibid.

49 Ibid; Thurston, "Vt. Yankee offers power deal on eve of critical vote."

50 Terri Hallenbeck, "Offer of cheap power won't delay VY vote," *Burlington Free Press,* February 24, 2010.

51 John Dillon, "On eve of senate vote, Entergy offers discounted power," Vermont Public Radio, February 23, 2010. See also Thurston, "Vt. Yankee offers power deal on eve of critical vote;" Hallenbeck, "Offer of cheap power won't delay VY vote."

52 Bob Stannard, email, February 23, 2010.

[53] Jess Edgerly, April 18, 2011; Kristin Carlson, "Vt. Senate votes to close Vt. Yankee," WCAX, February 24, 2010, http://www.wcax.com/story/12036106/vt-senate-votes-to-close-vt-yankee; Nancy Remsen, "Vermont Yankee: passion found on both sides of nuclear debate," *Burlington Free Press*, February 25, 2010.

[54] Center for Media and Democracy, "Under the dome: Vermont Senate vote on Vermont Yankee," February 24, 2010, http://www.cctv.org/watch-tv/programs/vermont-senate-vote-vermont-yankee; John Dillon, "Senate votes against Yankee re-licensing," Vermont Public Radio, February 24, 2010; Nancy Remsen, "Vermont Yankee: passion found on both sides of nuclear debate," *Burlington Free Press*, February 25, 2010.

[55] Jeanette White, June 11, 2011.

[56] Center for Media and Democracy, "Under the dome: Vermont Senate vote on Vermont Yankee."

SEVEN

Failure to Contain

T hroughout the spring of 2010, press stories continued to follow Entergy's efforts to locate and stop the tritium leak. Workers drilled additional wells, working around the clock to find the source. On March 21, one month after the Senate vote, Entergy announced the end of the massive 40-day operation. Investigators had found that the contaminated water had leaked from an underground concrete pipe tunnel. Two pipes housed in the tunnel contained radioactive fluids removed from the plant's cooling system. A drain hole to capture any fluids leaking from the pipes had become clogged with construction debris, dating back to 1978. As the pipes leaked and were unable to drain, the fluids slowly filled the concrete tunnel, eventually finding a crack where the pipes entered the tunnel and leaking out into the earth. Technicians flushed out 300,000 gallons of water from the area, removed silt and dirt and shipped the contaminated material to a waste site. In early April, Entergy invited reporters and legislators onto the site to view the cleanup efforts and meet with Entergy staff.

"We are embarrassed and disappointed by our performance and regret the issues this has caused," Mark Savoff, Entergy's executive vice president for operations, told a reporter. "We are taking the necessary actions to improve our performance and become a leader in tritium investigation, detection and remediation." Savoff said that Entergy would be working to rebuild Vermonters' confidence in the plant.[1]

The announcement provided a launching point for a new public relations campaign in early April. Full-page print ads apologized to Vermonters for the mistakes, took responsibility and promised to do better. The ads featured John Herron, president and CEO of Entergy Nuclear, the Entergy subsidiary that manages the company's nuclear reactors. A stern and unsmiling Herron, wearing a dark sports jacket and blue shirt, assured viewers that the plant's troubles were behind them: "I'm John Herron, president and CEO of Entergy Nuclear. I spent the first 15 years of my career right here at Vermont Yankee... I care very much about this plant and its role in providing jobs, taxes and electricity to Vermont. I understand Vermont values ... I recognize that we have an uphill battle to rebuild your trust but I know the hard-working men and women of Vermont Yankee are committed to doing just that ..."[2]

That spring, Entergy set up an "aggressive communications" campaign with government officials, regulators and utilities, which included daily emails and frequent phone calls. Led by the New Orleans-based Curt Hébert, the company worked hard to refurbish its image. Hébert and other Entergy staff members met with more than 100 stakeholders, taking responsibility, pledging to do better and asking for advice.

Entergy's supporters at the Department of Public Service and in the Governor's Office also renewed their commitment to the plant. DPS Commissioner David O'Brien said the decision on relicensing the plant was still open and the Senate's vote was "politics and nothing more than that." Governor Douglas added that the vote could be reversed by a future group of legislators, cautioning that votes should be based on the facts and political leaders should not "let emotion trump logic." Economists working for the state released a report concluding that a plant closure could cause 1,000 job losses and cost the state economy more than $60 million a year.[3] The union representing Vermont Yankee workers hosted a State House press conference, calling for the plant to remain open. The company's allied business groups promoted messages about jobs and low-cost energy.[4]

Entergy also continued the "I Am VY" ads, featuring business leaders concerned about the high price of energy if Yankee were to close. Based on interviews with television advertising departments,

Seven Days's Andy Bromage calculated the company's end-of-year 2009 and 2010 advertising spending at $500,000. The ads featured business leaders, plant employees and local citizens. For example, a St. Albans dairy farmer, with black-and-white milking cows in the barn behind him, said, "My biggest concern if Vermont Yankee were to close is that I haven't heard what the plan is for purchasing power to replace what they generate—and what the final cost is going to be to the ratepayer. It's not going to take too much more before a lot of us in the dairy industry can't continue."[5]

The town of Vernon issued a letter—"Vernon Speaks"—pointing out the substantial contributions of the plant to the local economy and local charities. More than 75 families in the small town had someone directly working at the plant. In March, a *New York Times* article, "Town finds good neighbor in nuclear plant," focused on the devastation the plant closing would have on the community. "It will ruin this town," a member of the selectboard told the reporter. In April, *USA Today* weighed in with a summary of the debate, stating that the issue remained open: "In Vermont, nuke power faces a test."[6]

In May, Hébert wrote his boss, Entergy CEO J. Wayne Leonard, that these efforts were having some success. One success Hébert pointed to was the lack of action by the state's House of Representatives. Following the Senate vote, activists had focused on a House vote, to further deepen the company's troubles. A second "no" vote would make it more difficult for a future Legislature to change course. For Entergy, another "no" vote would be yet another public rejection. The House did not take up the issue. "So far, so good," Hébert wrote.[7]

Political Landscape Changes

In May, Peter Shumlin, the State Senate's president, announced he was a candidate for the state's top job. Twelve days after the legislature adjourned, Shumlin stood with David Blittersdorf, a renewable-energy developer and leading Vermont Yankee opponent, and Beth Robinson, a lawyer and prominent gay-marriage advocate, to launch his campaign. With five candidates in the Democratic primary vying for the state's top job, many saw Shumlin's candidacy as a long shot. Furthermore, Douglas's chosen successor, Lt. Governor Brian Dubie, was a popular moderate who had beaten Shumlin in a previous election.

Shumlin touted his role in leading the fight to close Vermont Yankee and in winning passage of a law allowing gay marriage. A number of activist groups and donors, closely engaged in these separate efforts, saw Shumlin as the one candidate who had delivered results. Still, even with the support of these key blocks of likely Democratic primary voters, Shumlin's southern Vermont origins and lack of name recognition in Chittenden County—home to one-third of the state's primary voters—would present severe obstacles. Doug Racine, a Chittenden County senator who had served as lieutenant governor for five terms, was a strong candidate, as was Secretary of State Deb Markowitz, who had won statewide office six times. Separating himself from these candidates in a primary in which less than 75,000 people would vote would be a challenge for Shumlin.[8]

An extremely talented retail politician with the numbers of many of Vermont's political elite dialed into his cell phone, Shumlin campaigned relentlessly up and down the state through June and July. Practically living in Chittenden County, Shumlin seemed to be everywhere. The candidate skillfully relied on his vast rolodex and personal relationships, energized campaign style and personal finances to broaden his support. In Vermont, television ads are unusual in a primary campaign, because of the per-voter costs involved—akin to using a megaphone to talk to your dinner partner. In July, Shumlin started airing television ads, spending funds that his opponents did not match.[9] Shumlin touted his leadership on Vermont Yankee to further separate himself from his Democratic opponents.[10] A postcard mailed to Democratic voters featured the cooling tower photo, with water gushing from the collapsed structure. Shumlin's message was blazoned across the destroyed structure: "I led the way to retire Vermont Yankee on schedule …"[11]

In August, Shumlin narrowly won the Democratic gubernatorial primary, defeating Racine by less than 200 votes, a fraction of a percent of the 75,000 votes cast. Markowitz trailed Shumlin by less than 700 votes. A swing of a few hundred votes in either direction would have changed the outcome.[12] A civil recount delayed the general campaign for several weeks. During that time the five candidates campaigned together, focusing their fire on Lt. Governor Brian Dubie, the Republican nominee for governor.[13]

A well-known state figure, Dubie had served as lieutenant governor for eight years, defeating several strong Democrats during that time. A full-time airline pilot, Dubie was also a lieutenant colonel in the Air Force Reserve and a lifelong Vermont resident with an extensive family and personal networks. Vermont hosts a squadron of F-15s, which were scrambled to defend the country's skies after 9/11, and Dubie was at Ground Zero shortly afterwards. In contrast to Douglas, Dubie was more liberal on renewable-energy issues such as wind power and solar and had become a champion of "green" businesses. Straightforward and popular, Dubie was the kind of candidate who wins state office in Vermont. Like Douglas, Dubie had supported Vermont Yankee's continued operation.

Dubie's position on Vermont Yankee became a liability to his campaign, as Shumlin and independent groups of his supporters zeroed in. An organization called Green Mountain Future attacked Dubie's position on the plant. The organization's first television ad showed images of the plant with newspaper headlines about leaks and Entergy misstatements. An announcer said, "The state legislature votes overwhelmingly against relicensing … Brian Dubie says he would have voted FOR. Dubie wants Vermont Yankee open another 20 years …" Over a photo of Dubie with the plant behind him, the announcer asked, "Want Vermont Yankee open another 20 years? Tell Brian Dubie NO."[14] In mid-September the group launched a second ad, again blasting Dubie for his support for the plant. Organized as a 527, the organization was able to spend unlimited money without providing details on funding sources. Press accounts put early spending at above $125,000.[15]

As the general campaign continued, Shumlin repeatedly drew a sharp distinction between himself and Dubie on Vermont Yankee, calling for increased taxes on the plant's high-level radioactive waste, and for Entergy to accelerate their cleanup of the contaminated groundwater. "Brian will stand up for the stockholders of Entergy Louisiana instead of protecting the pocketbooks and health and safety of the people of the state of Vermont," Shumlin charged.[16]

Dubie, sounding defensive at times, said that the plant should continue to operate as long as it was "safe" and that decisions about the plant's future should be made by experts, not politicians. "My commitment to Vermonters is that science should drive the policy," Dubie said at one of the campaign's 13 debates. "The fact is, [in] the last legislative session ... politics drove the policy."[17]

When the ballots were counted, Shumlin had defeated Dubie by about 3,700 votes—less than 1 percent of the 200,000 votes cast—another very close election.[18] Two days later, Entergy announced that Vermont Yankee was for sale.[19]

Vermont Yankee Goes up for Sale

Rumors that the plant was for sale had been circulating for weeks. Following Dubie's announcement that he would not contest the election results, Entergy formally put the plant on the auction block. Citing the plant's high performance measures, recent operating record and long history, Entergy officials highlighted the plant's value. Entergy was joined by DPS Commissioner O'Brien, who said the plant was a safe and valuable asset that should be kept running, adding that the plant's "problems," while appearing frequently in the news "do not pose health concerns."[20]

Rep. Tony Klein was more blunt, calling the sale "an act of desperation," and adding, "Why would anybody buy that plant?"[21] VPIRG's Moore told a reporter, "Changing the out-of-state ownership to a new out-of-state corporation doesn't make the reactor one day younger. And that's its fundamental problem, it's falling apart. It's got a leak a week, and Vermont Yankee is past its prime."[22] Greenpeace parodied the announcement: "For Sale: Quaint Vermont fixer-upper from the last millennium. Nestled on the bank of the Connecticut River, this antique nuclear reactor features cozy relations with federal regulators and the new governor is from just down the road!"[23]

Figure 25. Photo of Vermont Yankee with "For Sale" sign. Courtesy of Greenpeace.

Yet Entergy had little choice. They needed the support of Vermont's two largest investor-owned utilities, Green Mountain Power and Central Vermont Public Service, to have any chance of success in Vermont.[24] To get that support they had to arrive at an agreement for the utilities to purchase plant electricity. The utilities had set four conditions before they would agree to buy Vermont Yankee's electricity, conditions that must have both infuriated and exhausted Entergy management. To win CVPS and GMP's support, the plant would have to have its NRC license in hand, have a purchase agreement for at least 20 megawatts of power from a third Vermont utility, resolve outstanding issues with the state *and* sell the plant to a new owner.[25]

The company pursued all of these fronts. The following spring, the plant's vice president for government affairs, Brian Cosgrove, said that they were close to announcing an agreement with the Vermont Electric Cooperative.[26] This would provide CVPS and GMP with added cover for a deal with Entergy. Unfortunately for Entergy, it would not be so simple. Early negotiations with the

Vermont Electric Co-op's general manager were rejected by the company's publicly elected board. The chastened general manager told a reporter that lack of trust with Entergy—despite the very favorable terms of the agreement—trumped everything else. Distrust was so high that he was not only reprimanded for negotiating with Entergy but also instructed to never speak to the company again. The general manager said that the lack of trust was "a sad thing," adding that he felt "really ... bad for the folks at Vermont Yankee. They are intelligent, hard workers. Yankee is a very well-run operation."[27]

NRC Reviews License

On March 10, two months after Shumlin was sworn in as governor and a year after the Vermont Senate voted to close the plant, NRC commissioners announced preliminary approval for the license extension.[28]

One day later, a massive earthquake triggered a giant tsunami, decimating Japan's northeastern coast and knocking out the electric grid feeding the three operating reactors at the Fukushima nuclear power plant. The world watched as Japanese operators struggled to provide desperately needed water to cool the reactors' cores, spraying sea water through hoses and water cannons and dumping it from helicopters. As the crisis unfolded, public officials asked for a pause on U.S. relicensing decisions until the implications for the country's nuclear fleet could be more fully understood. Reporters and advocates pointed out that the Fukushima reactors were the same design, year and vintage as Vermont Yankee—all built by General Electric in the early 1970s.

U.S. Senator Bernie Sanders (I-Vermont) called for a moratorium on license renewals, telling the Associated Press, "It's hard to understand how the NRC could move forward for a license extension for Vermont Yankee at exactly the same time as a nuclear reactor of similar design is in partial meltdown in Japan."[29]

Still, as expected, on March 21, 2011, the NRC issued the license, keeping alive a string of successes for nuclear power plant operators seeking extensions. To date, all 71 applicants that have completed the relicensing process have been approved.[30] Another

15 have applied and 17 more are expected to apply. This one took longer than usual, almost five years from when Entergy first submitted in January 2007. "We believe that Entergy, through the exhaustive review that we've done … meets all the requirements and standards to operate for another 20 years," NRC Chairman Gregory Jaczko told reporters in a conference call.[31]

Another Twist in the Story

A few weeks after receiving the NRC license, Entergy announced they had given up trying to sell the plant. "The plant's strong operating performance was attractive to potential buyers; the political uncertainty was not," said Entergy President Richard Smith.[32] Company CEO J. Wayne Leonard told a group of financial analysts that the company was running out of options. "We're being pushed into a bit of a corner here. And there's a point where there's a line in the sand and we've got to make a decision whether we're pushed any further or not."[33]

On Monday morning, April 18, Entergy filed a lawsuit against the State of Vermont, naming Governor Shumlin, PSB regulators and the attorney general as defendants. In a telephone conference call, Entergy's Smith read a prepared statement. At the VPIRG offices, advocates listened intently, crowded around a speakerphone in Paul Burns's office. The office was silent as Smith talked for 20 minutes, laying out the reasons for the lawsuit and claiming that the Legislature "changed the rules" and left the company with "no other choice."[34] At GMP and at the Governor's Office, staff put the final touches on an announcement to about purchasing replacement power from Hydro-Québec. Behind the scenes, GMP staff finalized plans to buy electricity from New Hampshire's Seabrook nuclear power plant.

"It is clear our disagreement with the state of Vermont on the scope of its authority over Vermont Yankee cannot be resolved … Putting this dispute before a federal judge is the appropriate and responsible way to resolve this disagreement," Smith said.[35]

"Entergy Sues Vermont," blazed the *Reformer* in big bold type. "Plant Owner Sues Vermont Over License for Reactor," ran the headline in the *New York Times*. In a sign of how national the conversation had become, the *Reformer*'s Audette quoted a nuclear policy expert from Greenpeace's Washington, D.C., office. Vermont's congressional delegation, Senator Sanders, Senator Leahy and Congressman Welch, issued a joint statement saying that Entergy "should respect and abide by Vermont's laws."[36]

Governor Shumlin responded in a press conference with Attorney General William Sorrell: "It is my job to follow that law. The decision has been made by the Legislature to close the plant as scheduled, and the law requires the Legislature to make that decision, as it should." Repeating themes from the campaign, Shumlin continued, "The fact is, we got a leak a week, the plant is old and tired, Entergy Louisiana can't be trusted and we all have to understand it was designed to be closed in 2012."[37]

The action returned the decision to a legal and technical process. It was now up to the skill of Entergy's lawyers and experts and the lawyers in Vermont's Attorney General's Office. Efforts by the New England Coalition and other advocacy groups to intervene in the case were rejected. Entergy was back on ground that might be more friendly. A decision would come from one person, 74-year-old U.S. District Judge J. Garvan Murtha, who had been brought out of retirement to hear the case.

Of immediate interest was the looming March 21, 2012, scheduled closure date. With no permit from the Public Service Board, Entergy would have to close the plant the following spring. Entergy requested an immediate stay, arguing that they had to decide now (summer 2011) whether or not to refuel the plant that fall. Nuclear plants are refueled every 18 months, an expensive and time-consuming procedure. Costs in this case were estimated at $92 million.[38] Entergy argued that to refuel the plant, without knowing whether they would continue to operate past March, put them in a position of "irreparable harm." The company requested an immediate injunction. Following a round of hearings in July, Judge Murtha denied the preliminary injunction, suggesting instead an expedited court process to resolve the case before March.

A week later, Entergy announced they would go ahead with the refueling, placing an expensive bet that the company would continue operating beyond March 2012.[39] That morning when I pulled the *Burlington Free Press* from my mailbox, I suddenly had the feeling that this case could go on for a very long time. The *Free Press* headlined the story, "Vermont Yankee Nuclear Plant To Refuel Despite Uncertainty." As CLF's Mark Sinclair had said earlier that summer, "Litigation is just a small cost of keeping the plant open ... If this thing drags out to the Supreme Court and takes 10 years, that's a perfect scenario for Entergy."[40] The conflict in Vermont had also started to affect Entergy's other operations, leading to a downgrade in the company's stock in June.[41]

Entergy did not confine their legal activities to Vermont, but also lobbied heavily for the NRC and the U.S. Justice Department to intervene on the company's behalf. In a neat piece of reporting, the Associated Press's David Gram found that a few weeks after filing the lawsuit in U.S. District Court, Entergy executives contacted officials at the Department of Justice, Department of Energy and the NRC, putting pressure on those officials to side with the company. Two Entergy organizations reported spending more than $760,000 on lobbying in D.C. from April to June of 2011.[42]

But all eyes would turn back to Brattleboro in mid-September when Judge Murtha gaveled in the first of three days of hearings. It was a beautiful fall day in Vermont, with a hint of color on the trees. After so many years and so much local debate, it seemed fitting that the case would be tried in Brattleboro, just a few miles from the plant and home to so much activity around the plant for the last 40 years.

Lawyers Debate the Plant's Future

Entergy's top-tier external counsel was led by Kathleen Sullivan, of the New York City law firm Quinn Emanuel Urquhart & Sullivan. Sullivan is a former constitutional law professor at Harvard, former dean of the Stanford Law School, a frequent litigator on constitutional issues before the U.S. Supreme Court and rumored to be on President Obama's short list for a seat on the U.S. Supreme Court. Sullivan's reputation and skills brought observers from around the state, just to hear her argue the case.

Entergy made two central arguments in the lawsuit. In the

first place, Entergy argued, Vermont's actions were preempted under federal law. Under the Atomic Energy Act, only the federal government can make decisions about the radiological safety of nuclear power plants. Safety was at the core of the legislative statutes in question; all the other issues were merely "pretexts," Entergy argued. Based on this preemption claim, Entergy requested that Act 160 be entirely stricken, along with sections of Act 74 (the dry-cask storage law from 2005) and Act 189 (that established the Legislature's oversight panel and the reliability audit). Entergy also stated that the Public Service Board did not have authority to regulate the plant, essentially suggesting that the 2002 agreement be voided. [43]

Secondly, Entergy argued that Vermont was interfering with interstate commerce by attempting to regulate cross-border electricity markets. Forcing the company to reduce Vermont's electric rates would have the effect of increasing rates charged in other states, violating interstate commerce laws. Because Vermont had requested a power agreement that benefitted state ratepayers, Vermont was essentially holding Entergy hostage. The company argued that, "Vermont expressly conditioned the approval of a renewed CPG—a coercive, regulatory measure—on VY agreeing to favorable, below-market rates for Vermont utilities."[44]

Vermont responded that states have broad rights to regulate and review issues—outside of safety —in nuclear power plants within state borders. The legislative laws under debate, Acts 160, 74 and 189, dealt with issues of reliability, economics, land use, operating capability and reputation—not safety—state lawyers argued. Peter Bradford, a member of the Legislature's oversight panel and a former NRC commissioner, served as a key state witness. In the 1970s, Bradford was a member of the NRC when Three Mile Island's partial meltdown triggered the events that have since become branded into public consciousness.[45] Bradford drew on his long career as a regulator at the NRC and in state agencies in Maine and New York to draw a picture of overlapping jurisdictions, of shaded relationships between state and federal regulators and of shared responsibilities.

Vermont's lawyers also pointed out that Entergy agreed to state regulation on multiple occasions, in the agreement they signed when they bought the plant, in statements when Act 160 passed and in letters to legislators and state regulators. If Entergy thought these laws were preempted, the state's lawyers argued, then why didn't they say that at some point over the previous eight years?[46] In fact, Entergy lawyers did the exact opposite. In 2002, Entergy told the PSB that their commitment to seek a state permit would not be preempted, and that states have the authority to consider "need, reliability, cost and other related state concerns." At the time, opponents to the plant's sale had argued that any agreement Entergy made would be suspect because of the preemption challenges. The Board rejected this argument, in part, because of the "binding contractual agreement" Entergy signed committing to seeking Board approval to operate beyond 2012.[47]

Both sides attached hundreds of pages of legislative hearing testimony, internal emails, newspaper articles and letters to illustrate their arguments. Examining the record from just one of the legislative statutes in question, Act 160, illustrates the "sausage-making" enterprise of legislating in full bloom. Over the course of the four-month legislative session, up to 16 different legislators spent 35 hours discussing the bill and taking testimony in different committee rooms. Entergy cited 42 excerpts from these discussions to tell a story of legislators, state officials and other participants concerned primarily about safety but weaving a "pretext" of other issues, such as reliability, trust, economic harm and decommissioning costs.

Relying on this legislative record, Entergy's Sullivan painted a picture of legislators focused on radiological safety. In her four-hour closing presentation on the third and final day of trial, Sullivan relied on an extensive PowerPoint to illustrate this argument. Sullivan would show the highlighted text in question, then click on it to play an audio of the legislator talking.[48] Sullivan played one legislator saying, "If we base our legislation on what we learned from our constituents most of that is going to be about safety." When told that safety was preempted, a legislator said, "Okay, let's find another word for safety." As they drafted the bill, a witness offered legislators advice in finding "language to help with preemption problems." [49]

Entergy used the same approach on the interstate commerce arguments, citing excerpts from the record to show that legislators and regulators attempted to force Entergy to lower the cost of the plant's electric output. Entergy quoted then-senator Shumlin as saying, "There's no way we're going to vote to relicense the plant unless Vermonters are getting a great deal."[50] Testimony from Dave Lamont, Vermont's director of regulated utility planning, also shows up in Entergy's brief: "The basis of the bargain should be that ratepayers are afforded a materially favorable power supply agreement in return for accepting certain risks that are unique to a nuclear facility" (see also Chapter 4). [51]

Lawyers for the state argued that Entergy cherry-picked the legislative record to "draw unfounded, sweeping conclusions of legislative pretext."[52] Holding 180 legislators accountable for the statements of a few ran counter to previous decisions, most notably the 1983 U.S. Supreme Court ruling in *Pacific Gas & Electric*. In that case the Supreme Court upheld the rights of states to review nuclear power plants in their issues of traditional jurisdiction, such as land use, reliability, water and air quality, economic factors and alignment with state energy plans. A free and spirited legislative debate takes place over many months in multiple forums with many people. The Supreme Court, recognizing this essential attribute, said that it would not try to understand what legislators were thinking, but instead review the approved legislative texts.[53]

The state also focused on Entergy's repeated commitment to abide by state laws and to seek a permit from the state. The 2002 Memorandum of Understanding was only one part of this argument. In the debate over the laws in question, Entergy never raised preemption issues, the state argued. The testimony of Entergy's Jay Thayer—who had been placed on administrative leave (see Chapter 5) and was now working for another Entergy subsidiary—featured prominently. In 2008 Thayer had said the state had the right to vote on Vermont Yankee, writing then, "This year, the Vermont legislature and the Vermont PSB will decide whether to allow for the continued operation of Vermont Yankee for another 20 years." Thayer was asked on the stand if the Legislature had a choice in deciding whether to allow Vermont Yankee to continue to operate. "That was the status in 2008, yes," said Thayer.[54]

Curt Hébert's efforts in Vermont in the spring of 2010 also became a key part of the state's case. In the months leading up to the Senate's vote, Hébert had been everywhere, meeting with reporters and stakeholders, hosting tense press conferences, dispensing his southern wisdom to increasingly skeptical reporters, "putting a face on a moose" as he described it the day before the Senate's vote. That spring Hébert had written his boss, Entergy CEO J. Wayne Leonard, a five-page memo summarizing his 2010 efforts to rehabilitate the plant. In stark language Hébert spelled out a number of "challenges" related to the plant's public meltdown: the tritium leak, Entergy's credibility, mistrust over the spin-off, lack of a power agreement with state utilities and problems with the decommissioning fund. Not once in his memo, Vermont's lawyers pointed out, did Hébert mention safety or radiological safety concerns as the state's secret motive in the rejecting Vermont Yankee.[55]

The Judge Rules

Throughout the fall, as the deadline to close the plant drew nearer, supporters and opponents waited for the judge's ruling. Finally, two weeks into the legislative session, late on a Thursday afternoon, Judge Murtha issued his 104-page decision. Murtha agreed with Entergy that the Vermont Legislature's 2006 law, Act 160, was based on radiological safety, as was a section of the 2005 law, Act 74, requiring Entergy to seek approval for storing waste at the plant. Murtha ordered those sections of the law struck down, voiding the Senate's 2010 vote.

At the same time, Murtha upheld the role of the state's Public Service Board, requiring that Entergy seek a state permit to operate in Vermont, citing an NRC document: "After the NRC makes its decisions based on the safety and environmental considerations, the final decision on whether or not to continue operating the nuclear power plant will be made by the utility, State and Federal (non-NRC) decision-makers. This final decision will be based on economics, energy reliability goals, and other objectives over which the other entities may have jurisdiction."[56] Murtha threw a confusing twist into

Figure 26. Members of the House and Senate energy bill conference committee discuss the proposed tax on Vermont Yankee to fund plant decommissioning. Facing are Representatives Shap Smith (D-Morristown), Robert Dostis (D-Waterbury) and Tony Klein, (D-East Montpelier). Photo by Glenn Russell, *Burlington Free Press.*

the question by potentially constraining the PSB, ordering that they not be allowed to require Entergy to sell electricity at a price "below-wholesale-market." [57]

Murtha drew heavily on Entergy's summary of the legislative debate, quoting from it extensively in the narrative part of the decision. More than half of the 104 pages were taken up with direct quotes from legislators, state officials and witnesses during legislative debates. For example, almost 12 pages were allocated to Act 160, as the judge took readers through every step of the bill, from its first hearing in the Senate Finance committee in February 2006, through other committee meetings and debates on the House and Senate floors. Not only legislators, but witnesses also were cited to show the Legislature's focus on radiological safety and low-priced electricity.

News media saw it as a victory for Entergy; no longer would the plant have to close in March. The *Free Press* ran a banner headline, "A stinging defeat," while also printing the iconic photo of the 2007 cooling tower collapse. Entergy released a statement: "The ruling is good news for our 600 employees, the environment and New England residents and industries that depend on clean, affordable, reliable power provided by Vermont Yankee."[58]

At the Capitol, Entergy's lobbyist said it was "a good day to be in the State House." At the VPIRG offices there was a pall in the air, but VPIRG's Director Paul Burns seemed energized, saying, "This is why we fight." Rep. Tony Klein was in his glass-walled committee room, right off the State House cafeteria, where the AP's David Gram had set up shop at a table, interviewing legislators and lobbyists. In the small lobby leading to the cafeteria, radio talk-show host Mark Johnson interviewed guests. Klein was particularly irritated at the chilling effect the judge's order would have on legislative debate and Vermont's open political culture. Vermonters' willingness to publicly debate issues and to work through issues in a public manner was under assault, Klein said.

Attorney General William Sorrell seemed to signal a likely appeal based on Murtha's focus on the law-making and not the law. Michael Dworkin, the former PSB chair, agreed: "Judge Murtha spent about two-thirds or three-quarters of his 100 pages looking at the process, looking at the sausage-making and we just think that's dead wrong." Like Klein, Dworkin was also concerned with the citations to witnesses as well as legislators: "This emphasis on what the witnesses said and what the Legislature was hearing is really a bad practice in terms of fundamental citizen involvement in a democratic government."[59]

State officials were also expected to challenge efforts to constrain the Board on the question of economic benefits. Like their counterparts in other states, state regulators have always looked at the value of utility contracts and the impacts on ratepayers when making decisions about state-hosted power plants. Murtha's focus on "below-market rates" in the narrative of the decision was seen as ambiguous because electricity rates, negotiated between a buyer and a seller, depend on the attributes of the contract (e.g. length of time, the amount of power purchased) and a host of other factors.

Consider the following different prices and contracts signed and proposed in Vermont in the last few years: In December 2009, Entergy offered a block of power to Vermont utilities at $.061/kWh. Two months later Entergy offered a smaller block at $.04/kWh. A few months later, they offered $.049/kWh for a 20-year contract. On the day Entergy filed its lawsuit, the PSB approved a 26-year, 225-megawatt deal starting at $.058/kWh between Vermont utilities and Hydro-Québec. And a few months before the lawyers tangled in Murtha's courtroom, Green Mountain Power announced an agreement to purchase 60 megawatts of electricity from New Hampshire's Seabrook reactor at a starting price of $.0466/kWh over 23 years.[60] These rates were set in negotiations between parties based on a whole host of factors, including their expectation of what the market might be in the future.[61]

State Oversight Continues

Despite the victory for Entergy in federal court, state oversight of Vermont Yankee will continue. Entergy's 2002 commitment to seek a state permit remains in effect. Regardless of court appeals, the PSB will need to issue a certificate of public good for the plant to operate in Vermont. Ten days after Judge Murtha's ruling, Entergy asked the Board for a speedy decision, saying the Board should have all the information it needs.[62] At the time of this writing, no timeline for a decision had been set. However, certain factors have clearly changed since the relicensing case went "on hold" back in 2009.

Vermont's electric utilities ended their Vermont Yankee power purchases in March. The state's Comprehensive Energy Plan—which develops, through a lengthy public process, a vision for the state's energy future—sees no role for Vermont Yankee in that future. Instead, the plan outlines moving all Vermont energy sources to 90 percent renewable energy by 2050, to promote economic security, create jobs and invest in Vermont communities.[63] After 40 years, electricity from the state's nuclear power plant is not included in that vision.

The state hosts a nuclear power plant that no longer provides electricity to state ratepayers. The plant employs Vermonters and provides tax revenues for state coffers, but is not part of the state's electric mix. Governor Shumlin said to a reporter a week after Judge Murtha's decision, "I think a lot of Vermonters are looking at the news right now going, 'What? This thing is going to keep running and we don't get any of the benefit?'"[64]

Entergy remains isolated politically and culturally, at odds with the state's political leaders and citizens. Two weeks after Judge Murtha's decision Entergy requested Vermont pay $4.6 million in legal fees to reimburse the company for the trial. At the same time, Vermont legislators were actively marking up bills to raise taxes on radioactive waste stored at the plant; the Agency of Natural Resources was reviewing the plant's water quality permit, and the Conservation Law Foundation continued with their appeal of the plant's NRC license because of the lack of a Clean Water Act permit. [65]

The state continues to have a variety of oversight responsibilities and clearly intends to exercise those responsibilities. When the Board denied Entergy's spin-off of Vermont Yankee, they wrote: "There are few more important public trusts than that assumed by the … owners of nuclear power plants to ensure their competent and reliable operation …" That is a public trust that regulators in Vermont take seriously and will certainly examine closely when hearings resume on the company's application for a certificate of public good.

Why and What Next?

At the beginning of this book I asked two questions;
1. How did a plant originally seen as certain to be relicensed move from being a core piece of Vermont's infrastructure to being rejected by Vermont's political leaders?
2. What is the role of states that host nuclear power plants in the decision-making that affects their citizens?

At the heart of the answer to the first question is the relationship between Entergy and Vermont's citizens and leaders. Over the ten years of Entergy's ownership of Vermont Yankee this relationship deteriorated. As GMP's Mary Powell said: "Every state has its character and every state has its qualities that you have to learn about. Particularly in small states, there's a really tight-knit fabric of relationships and there aren't that many policy leaders, both formal and informal. In Vermont, it's very important that you understand your operating environment. It's very important you deeply immerse yourself in the culture of this state."[66]

Electric decision-making in Vermont is an extremely public process. It was this very public debate that Entergy struggled with. The debate was about more than the cost of power, it was also about public trust, community engagement, transparency and the relationship of the company to Vermonters. Any company working in Vermont must operate on a foundation of trust both with its customers and with the communities that it serves. In Vermont, citizens and groups participate actively in the debate about where their energy comes from and the people and organizations that manage it. They voice their opinions to neighbors, in town meetings and to their political leaders. And they listen carefully.[67]

As for the question of what this story says about states' roles in the oversight of nuclear power plants within their borders, the answer remains undetermined. Clearly, there are a number of compelling non-safety issues of interest to states that host nuclear power plants, from the economic and environmental impacts, to the plant's role in the state's vision for its own energy future. These are important decisions for the state's citizens and political leaders.

The role of the states in nuclear power decision-making will continue to be debated. New York's Governor Andrew Cuomo and local community leaders oppose the present relicensing efforts of Entergy's two reactors at Indian Point—40 miles from New York City. Another Entergy plant, the Pilgrim reactor in Massachusetts, is also mired in a contentious relicensing process. Also in the spotlight is the renewal process for Diablo Canyon in California. A total of 15 license-extension applications are currently before the NRC and 17 more are expected. This first round of extensions for the very oldest plants will expire in the 2030s. Will there be another round of license extensions then? What will the role of states be in that possible future?

For Vermont Yankee, the subject of this particular story, the

future remains uncertain. As this book goes to press, the Vermont
Public Service Board will take up the case. Future court debates are
likely. The media will continue to cover the story closely, providing
rich details for the state's citizens to review the unfolding story.
Citizens and legislators will likely have plenty more to say. But
for now, the plant will continue to operate, selling its power into
wholesale markets, fully legal with its NRC license in hand.

Figure 27. Anglers admire their catch as they ice-fish on the Connecticut
River with the Vermont Yankee nuclear power plant behind them. Photo
by Glenn Russell, *Burlington Free Press*.

Notes

1   Bob Audette, "VY seeks to rebuild trust," *Brattleboro Reformer*, March
    25, 2010; David Gram, "Tritium leak apparently has stopped," Associated
    Press, March 21, 2010; Bob Audette, "Legislators visit leak site at VY plant,"
    Brattleboro Reformer, April 4, 2010. Note: Ironically on the same day, New
    York State regulators announced they were rejecting the Entergy spin-off plan,
    citing issues of trust raised in the Vermont case among other things ("NY rejects
    Entergy nuclear reactor spinoff plan," *Reuters*, March 25, 2010). In June 2010,
    the Vermont PSB also rejected the spin-off plan.
2   http://www.youtube.com/watch?v=iKaWSAWXp_A.
3   David Gram, "NRC reassures lawmakers about VY radioactive leaks,"
    Associated Press, April 1, 2010.
4   John Curran, "Workers say Vermont Yankee safe despite problems," Associated
    Press, September 27, 2010. Chris Garofolo, "Pro-VY firms urge Shumlin to let
    PSB do its work," *Brattleboro Reformer*, February 4, 2011. Curt Hébert, email
    to J. Wayne Leonard, May 6, 2010.
5   I Am Vermont Yankee.
6   Katie Zezima, "Town finds good neighbor in nuclear plant," *New York Times*,
    March 3, 2010, http://www.nytimes.com/2010/03/04/us/04vernon.html; Rich
    Hampson, "In Vermont, nuke power faces a test," *USA Today*, April 15, 2010;
    Chris Garofolo, "Vernon still backs Yankee," *Brattleboro Reformer*, February
    24, 2010.
7   Curt Hébert, email to J. Wayne Leonard, May 6, 2010; Paul Burns, in discussion
    with the author, June 1, 2011. Note: Shap Smith later testified in the court case
    that there was no reason for the House to vote; since the Senate had voted,
    the issue was effectively null. Smith also testified about Entergy requesting
    the House not vote, testimony that was used as evidence to illustrate Entergy
    thought the vote had significance as the memo had suggested, at odds with their
    position in the lawsuit.
8   Lisa Rathke, "Shumlin kicks off campaign for governor," Associated Press,
    May 24, 2010.
9   http://vermont-elections.org/elections1/campaign_finance_2010_filings.
    html. Note: In terms of the primary, which took place on August 25, 2010,
    total spending up to August 17 was: $521,375 for Shumlin, $65,533 for Susan
    Bartlett, $246,223 for Matt Dunne, $576,433 for Deb Markowitz, $210,708 for
    Doug Racine.
10 David Gram, "With crowded field, Vt. gov primary unpredictable," Associated
    Press, August 22, 2010.
11 Shumlin for Governor, campaign postcard.
12 David Gram and John Curran, "A top Vt. senator claims victory in governor
    race," Associated Press, August 25, 2010.
13 Shumlin would later name both Doug Racine and Deb Markowitz to positions
    in his cabinet.
14 Text of ad.
15 John Curran, "Group targets GOP gov nominee over Vermont Yankee,"
    Associated Press, September 21, 2010.

[16] David Gram, "Vt.'s Shumlin: Tax high-level radioactive waste," Associated Press, October 1, 2010; David Gram, "Shumlin: Vermont nuke plant should step up cleanup," Associated Press, October 11, 2010.

[17] David Gram, "Vt. gov candidates spar over future of nuke plant," Associated Press, October 6, 2010.

[18] United Press International, "Democrat Shumlin next Vermont governor," November 3, 2010, http://www.upi.com/Top_News/US/2010/11/03/Democrat-Shumlin-next-Vermont-governor/UPI-87781288822825/.

[19] David Gram, "Vt. nuke plant for sale; gov.-elect not impressed," Associated Press, November 4, 2010.

[20] Bob Audette, "Vt. DPS chief: VY is safe," *Brattleboro Reformer*, October 21, 2010.

[21] David Gram, "Entergy to put VY up for sale," Associated Press, November 4, 2010.

[22] John Dillon, "Reports say that Entergy wants to sell Vermont Yankee," Vermont Public Radio, October 25, 2010, http://www.vpr.net/news_detail/89104/reports-say-that-entergy-wants-to-sell-vermont-yan/ (accessed December 21, 2010).

[23] Greenpeace. http://www.vermontyankee4sale.com.

[24] Curt Hébert, email to J. Wayne Leonard, May 6, 2010; Robert Young, in conversation with the author, August 24, 2011.

[25] Bob Audette, "Entergy unable to sell VY," *Brattleboro Reformer*, March 31, 2011.

[26] VEC had its own troubled nuclear power history; the rural electric cooperative was forced into bankruptcy because of its investments in Seabrook's nuclear reactor in the 1980s.

[27] Bob Audette, "VEC rejects VY power buy offer: Cites distrust of Entergy as reason for the decision," *Brattleboro Reformer*, April 27, 2011; http://www.vermontelectric.coop/news-center/211-vec-board-rejects-vermont-yankee-offer.

[28] David Gram, "Vermont Yankee gets federal license renewal," Associated Press, March 10, 2011

[29] David Gram, "Despite calls to slow down, NRC grants Vt. renewal," Associated Press, March 21, 2011.

[30] Associated Press, "NRC, nuclear industry rewrite history," June 28, 2011; http://www.huffingtonpost.com/2011/06/28/nrc-nuclear-industry-ap-investigation_n_885862.html.

[31] David Gram, "Vermont Yankee gets federal license renewal," Associated Press, March 10, 2011. Note: In the LILCO case of Shoreham on Long Island, the NRC issued a license after the plant had agreed to close. In that case, NRC may have been trying to assert their authority. See book by Joan B. Aron, *Licensed to Kill? The Nuclear Regulatory Commission and the Shoreham Power Plant* (Pittsburgh, PA: University of Pittsburgh Press, 1998).

[32] Bob Audette, "Entergy unable to sell VY," *Brattleboro Reformer*, March 31, 2011.

[33] "Entergy boss points to federal say on nuclear plant," *Bloomberg*, February 2, 2009, http://www.bloomberg.com/news/2011-02-09/entergy-boss-points-to-federal-say-on-nuke-plant.html.

34 Matthew L. Wald, "Plant owner sues Vermont over license for reactor," *New York Times*, April 18, 2011; Bob Audette, "Entergy sues Vermont," *Brattleboro Reformer*, April 19, 2011.
35 Rick Smith, conference call remarks, April 18, 2011, http://www.entergy.com/global/VY/Rick_Smith_REMARKS_VY.pdf.
36 Wald, "Plant owner sues Vermont over license for reactor," Bob Audette, "VY to get new license next week," *Brattleboro Reformer*, March 11, 2011; David Gram, "Vermont Yankee gets federal license renewal," Associated Press, March 10, 2011.
37 Chris Garofolo, "Pro-VY firms urge Shumlin to let PSB do its work," *Brattleboro Reformer*, February 4, 2011; John Curran, "Entergy sues to keep Vermont nuclear plant open, Associated Press, April 18, 2011.
38 Anne Galloway, "Entergy to Refuel Vermont Yankee," *VTDigger.org*, July 25, 2011, http://vtdigger.org/2011/07/25/entergy-to-refuel-vermont-yankee/.
39 Associated Press, "Vt. nuclear plant to refuel despite uncertainty," July 25, 2011.
40 Mark Sinclair, in discussion with the author, June 1, 2011.
41 Terri Hallenbeck, "S&P downgrades credit rating for Entergy Corp," *Burlington Free Press*, June 29, 2011.
42 David Gram, "Documents show heavy Entergy lobbying on Vt. nuke," Associated Press, November 6, 2011.
43 Attorneys for Plaintiffs Entergy Nuclear Vermont Yankee, LLC and Entergy Nuclear Operations, Inc., "Complaint for Declaratory and Injunctive Relief," p. 24; http://www.atg.state.vt.us/assets/files/The%20State%20of%20Vermonts%20Answer%20to%20the%20Complaint%20for%20Declaratory%20and%20Injunctive%20Relief.pdf (accessed December 23, 2011).
44 Attorneys for Plaintiffs Entergy Nuclear Vermont Yankee, LLC and Entergy Nuclear Operations, Inc., "Plaintiffs' Post-Trial Brief," p. 17; http://www.atg.state.vt.us/assets/files/ENVY%20Post-Trial%20Brief.pdf (accessed December 23, 2011).
45 Sam Walker, *Three Mile Island: A nuclear crisis in historical perspective* (Berkeley: University of California Press, 2004).
46 Olga Peters, "State rests its case in Entergy lawsuit," VTDigger.org, September 14, 2011, http://vtdigger.org/2011/09/14/state-rests-its-case-in-entergy-lawsuit/ (accessed December 22, 2011).
47 Attorneys for Defendants, "Defendants' Post-Trial Brief," pp. 10-14; Michael Dworkin, in discussion with the author, January 19, 2012.
48 John Dillon, Vermont This Week, January 27, 2012.
49 Attorneys for Plaintiffs Entergy Nuclear Vermont Yankee, LLC and Entergy Nuclear Operations, Inc., "Plaintiffs' Post-Trial Brief," pp. 11-15.
50 Ibid, p. 17.
51 Vermont PSB, "Prefiled direct testimony of David Lamont," February 11, 2009, http://www.atg.state.vt.us/assets/files/CLF%20Exhibit%2015B.pdf, pp. 21-24.
52 Attorneys for Defendants, "Defendants' Post-Trial Brief," p. 10.
53 Attorneys for Defendants, "Defendants' Post-Trial Brief," pp. 10-14; Michael Dworkin, in discussion with the author, January 19, 2012.

[54] David Gram, "Nuke plant VP says firm agreed to Vt. oversight," Associated Press, September 13, 2011.

[55] Attorneys for Defendants, "Defendants' Post-Trial Brief," p. 12. Note: Hébert was no longer with Entergy at the time of the court case, with the reasons he left the company unclear, see Dan Barlow, "Shuffle at the top for Yankee," *Times Argus*, June 9, 2010, http://www.vermonttoday.com/apps/pbcs.dll/article?AID=/BT/20100609/NEWS01/6090353.

[56] Murtha decision, p. 59, from 61 Fed Reg. 28467, 28473, (June 5, 1996)(10 C.F.R. Part 51).

[57] Murtha, p. 101.

[58] Terri Hallenbeck, "The Yankee Challenge," *Burlington Free Press*, January 29, 2012.

[59] Vermont Edition, Michael Dworkin, William Sorrell, January 23, 2012; Jane Lindholm.

[60] Nancy Remsen, "Board approves power deal with Hydro-Québec," *Burlington Free Press*, April 18, 2011.

[61] Terri Hallenbeck, "Cheaper power is still nuclear," *Burlington Free Press*, May 25, 2011.

[62] John Dillon. "Entergy wants prompt OK to operate for 20 more years," Vermont Public Radio, January 31, 2012.

[63] 2011 Vermont Comprehensive Energy Plan. http://www.vtenergyplan.vermont.gov/sites/cep/files/CEP%20Overview%20Page_Final.pdf.

[64] Vermont Edition, Michael Dworkin, William Sorrell, January 23, 2012; Jane Lindholm; Tape played of Governor Shumlin from January 23, 2012.

[65] Terri Hallenbeck, "The Yankee Challenge," *Burlington Free Press*, January 29, 2012, "Vt. Yankee owner wants $4.6 million in legal fees," *Rutland Herald*. February 3, 2012.

[66] Mary Powell, in conversation with the author, July 5, 2011.

[67] Steve Terry, email, February 1, 2012. Note: The word "public" is the 8th most common word in the 1,409 news articles, showing up 3,521 times.

Acknowledgements

Writing a book with controversy at its heart and a stage full of different players is a journey in detail and careful word choice. To get it right I had the good fortune of receiving advice and direction from many friends and colleagues.

I could not have completed this project without the thoughtful feedback and superlative editing of my mentor and friend Stephanie Kaza. Her clear vision and encouragement have been invaluable. Stephanie is a talented writer, a superb editor and a gifted teacher. Thank you!

My gratitude also goes to Cheryl Morse, director of the Center for Research on Vermont, whose many sharp insights, support and willingness to step into the unknown opened the door for this project. Cheryl put together an editorial review committee and I thank those members for their advice along the way; Frank Bryan, Michael Dworkin, Stephen Terry and James Throgmorton. Steve and Michael's vast knowledge of energy planning in Vermont provided critical insights at key times. I've also depended on Frank's understanding of Vermont and Jim's thoughtful analysis of the nuclear industry.

Many friends and former energy-planning colleagues pitched in. Thanks go to Paul Burns, David Lamont, Sandra Levine, Richard Saudek and Ben Walsh for their keen suggestions. And special thanks to Mark Floegel for his early encouragement and enthusiasm. Jarred Cobb and Anne Bliss provided superb editorial assistance and Colleen Thomas added invaluable photo coordination. Glenn Russell's great photos, Susan McClellan's superb cover and Sue Storey's design took the book from words to life.

Past and present students at the University of Vermont, including Ben Carlson, Jonathan Dowds, Stefan Lutter, Mikayla McDonald and Phoebe Spencer all contributed invaluably. Jon Maddison, once a student and now a colleague, helped me think through data collection and analyses and the larger questions this book examines. A previous collaboration with Paul Hines was instrumental in guiding my thinking.

I offer special thanks to the more than 30 people who agreed to sit down and tell their story, trusting me and my tape recorder, and reminding me why we live in Vermont. I deeply appreciate the opportunity to tour Vermont Yankee, my guide Brian Cosgrove and the employees from the company who joined us for lunch to explain their work.

I also thank Jane Kolodinsky and my friends at CDAE for creating an environment where crazy ideas are welcome and people are given a chance to thrive.

Crea Lintilhac and David Blittersdorf, who believe in the power of ideas to change the world, provided early support that enabled the data collection and analysis that is at the core of this book. Thank you.

And lastly, thanks to my wife Allison and daughters Anna, Rose and Kristina, who kept life balanced along the journey and cheered me on through countless hours holed up on the computer. Allison, the journey started in a classroom and you've been schooling me ever since!

Richard Watts, February 3, 2012

CPSIA information can be obtained at www.ICGtesting.com
Printed in the USA
LVOW130136181212

312142LV00008B/387/P